History of Modern Psychology

A list of Jung's works appears at the back of the volume.

History of Modern Psychology

Lectures Delivered at ETH Zurich

Volume 1, 1933–1934

C. G. JUNG

EDITED BY ERNST FALZEDER

Foreword by Ulrich Hoerni

Translated by Mark Kyburz, John Peck, and Ernst Falzeder

PHILEMON SERIES

Published with the support of the Philemon Foundation
This book is part of the Philemon Series of the Philemon Foundation

Princeton University Press
Princeton and Oxford

Published by Princeton University Press

41 William Street, Princeton, New Jersey 08540

6 Oxford Street, Woodstock, Oxfordshire OX20 1TR

press.princeton.edu

Library of Congress Control Number: 2018954402

ISBN: 978-0-691-18169-1

British Library Cataloging-in-Publication Data is available

Editorial: Fred Appel and Thalia Leaf

Production Editorial: Karen Carter

Jacket Design: Kathleen Lynch / Black Kat Design

Jacket Credit: Laszlo Moholy-Nagy, "Am 7 (26)," 1926.

Production: Erin Suydam

Publicity: Tayler Lord

Copyeditor: Jay Boggis

This book has been composed in Sabon LT Std

Printed on acid-free paper. ∞

Printed in the United States of America

10 9 8 7 6 5 4 3 2 1

Contents

Foreword: C. G. Jung's Activities at ETH Zurich

ULRICH HOERNI

Translated by Heather McCartney

General Overview

In May 1933 C. G. Jung applied to the Swiss education board to be accepted as a lecturer in the field of modern psychology at the Swiss Federal Institute of Technology (Eidgenössische Technische Hochschule; henceforth: ETH).[1] Jung wrote that he wanted to resume his public lecturing, and would like to lecture on modern psychology, and was applying to ETH, as the topic could not be confined to the medical faculty. He requested ETH to recognize his status, gained at the University of Zurich in 1904, as a university lecturer (*Privatdocent*). Endorsed by evidently favorable references from ETH Professors Fritz Medicus (philosophy and pedagogy), and Eugen Böhler (economics) on 24 June 1933 the education board resolved to "grant Dr. Jung permission to publicize and hold lectures in psychology in the capacity of private lecturer in the general department for elective subjects at the ETH" and to award him a license to teach for eight semesters.[2] On 20 October 1933, Jung started his teaching activities, which continued until summer 1941. In 1935, he was awarded the title of "titulary" professor (*Titularprofessor*) by the Swiss government (the Federal Council).[3] He lectured for a total of thirteen semesters (he was on sabbatical for a further three).[4] During this

[1] Letter from Jung to Prof. Dr. Arthur Rohn, President of the Swiss Education Board, dated 2 May 1933 (Jung family archive). Prof. Dr. H. K. Fierz prompted Jung's application. (Personal communication from Prof. Dr. C. A. Meier, 10 February 1994). Jung and Fierz had traveled together to the Middle East in March 1933.

[2] Minutes of the Swiss Education Board Meeting of 24 June 1933 (Jung family archive) and letter to Jung from the Swiss Education Board, dated 24 June 1933 (ETH Archive). For Jung's correspondence with Böhler, see *C.G. Jung und Eugen Böhler: eine Begegnung in Briefen*, with an introduction by Gerhard Wehr, Zurich, Hochschulverlag an der ETH, 1996.

[3] Letter from Swiss Education Board to Jung, dated 26 January 1935.

[4] ETH Course Prospectus (ETH archive).

period, he introduced the totality of his theories, hypotheses, and methods to his audience; in fact, he presented his life's work as it then stood, something that cannot be found anywhere else in his writings. Alongside the *lectures* for a larger audience, from spring 1935 Jung also gave regular *seminars* for a small circle of participants. Unlike a part of the ETH Seminars,[5] the lectures have not as yet been published. This omission has now been rectified with this new publication.

The ETH Context

ETH[6] Zurich is a foundation of the Swiss Confederation.[7] Since the late Middle Ages, Switzerland had been a loose federation of small sovereign states without any superordinate state institutions (i.e., having no common government, capital city, currency, official language etc.). Following the Reformation, religious differences began to cripple harmonious coexistence. After a political crisis in 1847,[8] forces prevailed that sought to transform this state of affairs through the creation of a modern federal state. The constitution of 1848 contained a clause stating that the federal state was authorized to establish a national university, including a polytechnic. The development of technology (the railroad, industrialization, etc.) had created the need for such an institution, models of which already existed in France in Paris (since 1794) and in Germany in Karlsruhe (since 1826). In 1851, the Swiss parliament began drafting a bill to this end. However, conflicting internal

[5] C. G. *Jung, Seminare: Kinderträume*. Ed. Lorenz Jung and Maria Meyer-Grass (Olten: Walter, 1987). English edition: *Children's Dreams: Notes from the Seminar Given in 1936–1940*. Ed. Maria Meyer-Grass and Lorenz Jung. Trans. Ernst Falzeder with the collaboration of Tony Woolfson. Philemon Series (Princeton, NJ: Princeton University Press, 2010); and *Dream Interpretation: Ancient and Modern: Notes from the Seminar Given in 1936–1940*. Ed. John Peck, Maria Meyer-Grass, and Lorenz Jung. Trans. Ernst Falzeder with the collaboration of Tony Woolfson. Philemon Series (Princeton, NJ: Princeton University Press, 2014). Further seminars on association experiments from the same period have not as yet been published due to insufficient documentation.

[6] Eidgenössische Technische Hochschule Zürich / École polytechnique fédérale de Zurich / Politecnico federale di Zurigo / Swiss Federal Institute of Technology Zurich.

[7] The following sections on the history of the ETH are based mainly on two publications by the ETH: a) Various authors, *Eidgenössische Technische Hochschule 1855–1955* (Zurich: Neue Zürcher Zeitung Press, 1955); b) Jean-François Bergier and Hans Werner Tobler (eds.), *Eidgenössische Technische Hochschule 1955–1980, Festschrift zum 125-jährigen Bestehen* (Zurich: Neue Zürcher Zeitung Press, 1980). The developments of the ETH after Jung's retirement in 1941 and of the ETH Lausanne in 1969 are not covered in this synopsis.

[8] The "Sonderbund" war. The Sonderbund was a separate league of cantons within the Swiss Federation.

political interests posed difficulties. There was concern that the canton in which this institution would be built would possess too much power in the state. It was argued that the national university should therefore not be located in the town that would later be designated as the capital city.[9] It was also deemed important to consider the different constituencies in the country, so it was contended that the university should be built in the German-speaking area of the country, while the polytechnic should be situated where French was spoken. Francophone cantons cautioned against the negative cultural influences of an institution with a Swiss-German character (against "Germanification"), while Catholic cantons feared the negative consequences of an institute in a Protestant area. Further, there were financial considerations: the new state had as yet scarcely any income at its disposal. The cantons in question would be obliged to provide buildings for the respective institutions. There were already several universities in the country (in Basle, Zurich, Bern, Lausanne, and Geneva). The National Council (great chamber) finally approved the joint project in 1854. The council of states (small chamber) failed to ratify the university proposal, authorizing only a polytechnic. It apparently did not consider such an institution to be significant. The canton of Zurich expressed an interest in adopting the project; soon a federal polytechnic was finally created in the city of Zurich.

Courses at ETH

Courses began as early as the autumn of 1855. There were six departments: I. architecture; II. engineering; III. mechanical engineering; IV. chemical engineering; V. forestry; VI. department of philosophy and public economics (= elective subjects), including a) sciences; b) mathematics; c) literature and public economics; d) the arts. The curriculum was to be delivered in one of the national languages: German, French, or Italian. Even today at ETH, key subjects are taught in two or three of the national languages. Departments I–V offered solid courses leading to diplomas in technical disciplines. Admission to courses required a specific high-school qualification or an entrance examination. Department VI was to foster a grounding in general cultural values, without a specific final qualification. The subdepartments c) and d) were also open to members of the public wishing to attend single lectures, with no special entrance requirements.

[9] Bern was designated as the capital city.

In time, the departments at ETH increased, with some being reorganized or redesignated, although the basic structure remained the same. Traditional university disciplines such as theology, medicine, classics, and so on were not offered.

The Elective Subjects Department

Department VI c) was particularly important: It offered a broad spectrum of lectures in philosophical, literary, historical, and political education. The concept of the university lived on in this department, and in this way students were introduced to the culture of their fellow citizens from other language communities. It was obligatory to attend at least one such lecture per week. These subjects were taught only to a level of relevance to aspiring engineers and scientists. The teaching of subjects with a low requirement for hours of attendance was delivered not by permanently employed professors, but by private associate professors. The education board pursued a high quality of teaching and advertised assistant professorship appointments internationally. The German author and philosopher Friedrich Theodor Vischer and the Swiss art historian Jacob Burckhardt were among those appointed in department VI c).

The Building

At its opening in 1855, the polytechnic was not as yet housed in its own building. Since the canton of Zurich was required to make premises available, the courses initially took place in rooms belonging to the university near the present-day Bahnhofstrasse and in the canton's grammar school. However, the success of the new institute soon gave rise to a shortage of space. The notable German architect Gottfried Semper (recommended by Richard Wagner who lived in Zurich at the time) had been head of department I since 1855. He was commissioned to design a new building for ETH. Switzerland at that time had a population of ca. 2.5 million, the canton of Zurich ca. 270,000, the city of Zurich ca. 20,000.[10] The new polytechnic[11] was located in an area where the ancient city walls had recently been demolished. It cost a hefty sum for the time and was one of the largest buildings in Switzerland. It was opened in 1864 and included

[10] *Statistical Handbook of the Canton of Zurich* (Zurich: Office of Statistics of the canton of Zurich, 1949).

[11] On the history of the building, see Martin Fröhlich, "Semper's Main Building of the ETH Zurich," *Swiss Art Guide* (Bern: Society for Swiss Art History, 1990).

an astronomical observatory. It was situated on the edge of the fields and meadows of the outlying villages, which would later become residential areas of the city. The university (of Zurich) also required additional space, so it occupied the entire south wing of the new building. (The university's current main building was first opened in 1914.) The main building of ETH gradually acquired its present external form through large extensions and renovations that took place between 1915 and 1924. The façade of the north wing flaunts allegorical wall paintings, designed to proclaim both the authority of the Swiss federal state, represented by the armorial crests of the cantons, and the academic ambitions of the institute:

> NON FUERAT NASCI NISI AD HAS SCIENTIAE ARTES. HARUM PALMAM FERETIS
>
> *To be born would avail nothing if it were not for these sciences and arts. Through them you will achieve the victor's prize*[12]
>
> NUMINE, INDOLE, COGNOSCENDO, INTUENDO, MEDITATIONE, EXPERIMENTO, CONSTANTIA, IMPETU, EXEMPLO, INVENTIONE, ACUMINE, LABORE, DISCIPLINA, LIBERTATE, AUDACIA, CURA
>
> *By means of divine providence, giftedness, knowledge, contemplation, reflection, experiment, constancy, zeal, by example, invention, ingenuity, labor, discipline, liberty, boldness, diligence*
>
> SIMON LAPLACE, GEORGES CUVIER, CONRAD GESSNER, ALEXANDER VON HUMBOLDT, ISAAC NEWTON, LEONARDO DA VINCI, ARISTOTELES, PERIKLES, MICHELANGELO, ALBRECHT DÜRER, DANIEL BERNOULLI, GALILEO GALILEI, RAFAEL SANZIO, JACOB BERZELIUS, JAMES WATT

The Development of the Polytechnic

Although originally initiated through "small-state" politics, the polytechnic was an international success. Courses had begun in 1855 with 71 students. The collection of the holdings of the present-day ETH

[12] This quote refers to a group of female allegorical figures on the façade, representing various sciences and arts.

library began at the same time. In 1870, the copper engraving collection (today, the collection of prints and drawings), originally curated for history of art courses, was opened to the public. Women were admitted as students from 1871. In 1900, there were approximately 1000[13] students enrolled at the polytechnic. Of them, around two-thirds came from Switzerland, one-third from abroad, specifically from Austria-Hungary, Russia, and Germany. Even Albert Einstein studied at what was then the Institute for Specialist Teachers of Math and Science between 1898 and 1902. In 1905, the polytechnic was renamed the Swiss Federal Institute of Technology (ETH). At the time of Jung's application it included twelve departments with a total of approximately 1700 students and several hundred visiting students. The Department of Philosophy and Public Economics was now called Department XII for Elective Subjects and included philosophy/economics and math/science technical sections. The former was further divided into the areas of 1) literature, languages, philosophy (where Jung was active); and 2) historical and political sciences.

Psychology at ETH

The subject of psychology had already been established for some time when Jung arrived at ETH.[14] Between 1893/1894 and 1906, Professor August Stadler (1850–1910), a Kantian philosopher who had studied with Hermann von Helmholtz, had given lectures on psychology, alternating with philosophical subjects. From 1904, in conjunction with the university and on its campus, ETH offered lectures by Dr. Arthur Wreschner (1866–1932), the head of the psychology laboratory there in the faculty of philosophy and pedagogy. Wreschner qualified in philosophy and medicine. He conducted experiments on memory, the association of ideas, and the child's acquisition of language. Together with an ongoing "Introduction to Experimental Psychology," the lecture programs in those years included courses that one might not expect at a technical institute, such as "The Psychology of Feeling," "The Basic Facts of the Life of the Psyche," "The Psyche of the Child," "Feeling and Intellect," and "Psychology and Education." In addition to his experimental interests, Wreschner promoted applied psychology. Wreschner died in 1932. Jung's application in 1933 may well be seen

[13] Source of statistical data: ETH Course prospectuses 1855–1932 (ETH archive).

[14] ETH Course prospectuses 1855–1932 (ETH archive).

in the context of the vacancy that arose from his death. Nothing is known about ETH guidelines regarding the content of Jung's teaching activities. In his application, he had stated that he wished to lecture on the general topic of modern psychology. Since he would be working as a private assistant professor, in 1934 he established a fund, "The Psychology Fund," which was designed to support a private assistant professorship or a lectureship in general psychology. In the Fund's statutes,[15] there is a statement of purpose that reads as a general description of Jung's intentions:

> *The approach to this psychology should in general be determined by the principle of universality, i.e., no special theory nor special subject should be represented, rather psychology should be taught in its biological, ethnic, medical, philosophical, cultural-historical, and religious aspects. It is the purpose of this policy to liberate the teaching of the human psyche from the narrow confines of the discipline and to give the student, burdened by his professional studies, an overview and synopsis that facilitate an engagement in those areas of life that his professional studies do not cover. The lectures within the framework of general psychology should communicate to the student the possibility of a culture of the psyche.*

This statement gives evidence of the breadth of Jung's undertaking. It is also noteworthy that he did not see the fund as simply a vehicle to promote his own school of psychology. Jung's schedule at ETH comprised a weekly hour-long lecture (Fridays 6:00 pm to 7:00 pm) and, from 1935, a two-hour seminar,[16] in conjunction with his assistant, Dr. med. C. A. Meier (who himself later became professor of psychology at ETH). The language of the course was German. On average around two hundred and fifty[17] people registered for the lectures in the winter semester, around one hundred and fifty in the summer,[18] and around thirty for the seminars. The lectures were open to the general public; the seminars ("privatissime") to participants with adequate specialized background knowledge: "Pupils (i.e., Jung's training analysands), doctors, educationalists."[19]

---

[15] "Fund for the Promotion of Analytical Psychology and Related Subjects (Psychology Fund)," deed of donation, 15 September, 1934 (ETH archive).

[16] ETH Course prospectuses 1933–1941 (ETH archive).

[17] The winter semester usually ran from the end of October until the start of March, the summer semester from the end of April until the start of July.

[18] Statement of fees paid to C. G. Jung from the ETH 1933–1941 (Jung family archive).

[19] Letter from Jung to ETH, dated 19 July 1937 (Jung family archive).

ETH and the University

Between 1905 and 1913, Jung had lectured at the University of Zurich on psychiatry, hysteria, and psychopathology as well as on psychoanalysis and the psychology of the unconscious (from 1910). He made reference to this earlier teaching in both his application[20] and in his first ETH lecture.[21] However, a seamless continuation of those clinical lectures designed for the training of *doctors* was not what was required in 1933. ETH trained neither doctors nor psychologists. ETH and the university are very different institutions. The university was founded in 1833 by the canton of Zurich. It includes a number of traditional faculties, such as theology, law, and medicine. The Zurich University Hospital and the Burghölzli University Psychiatric Hospital are linked to the faculty of medicine and provide treatment for patients as well as teaching and research at the university. Admission to study demands specific entrance requirements. From 1901 to 1909 Jung worked as an assistant and resident doctor at the Burghölzli clinic, and he had begun his teaching activities while in this position.[22] To date, some notes of Jung's lectures in 1912–1913, taken by Fanny Bowditch Katz, have come to light.[23] A later point of contact arose when, in 1938, Jung and a group of colleagues of different orientations planned the founding of a training institute for psychotherapy. The university supported the plan, but the superior authority, the canton of Zurich board of education, refused the application to use university premises. Following this, Jung turned to ETH with the same request.[24] However, the project found no approval here with the explanation that ETH would provoke the university's displeasure if it encroached upon the university's subject areas.[25] The training institute for psychotherapy came to nothing.

The Significance of ETH for C. G. Jung

In the winter semester of 1941/42, Jung took sick leave from ETH,[26] and subsequently retired completely from his work there.[27] This was

---

[20] Cf. footnote 1.

[21] Lecture of 20 October 1933.

[22] Course prospectuses of the University of Zurich 1905–1913 (State Archives of the Canton of Zurich).

[23] Fanny Bowdith Katz papers, Countway Library for the History of Medicine, Boston.

[24] Letter from Jung to the ETH, dated 6 May 1939 (Jung family archive).

[25] Letter from the ETH to Jung, dated 17 May 1939 (Jung family archive).

[26] Minutes of the Swiss Education Board of 2 September 1941.

[27] Minutes of the Swiss Education Board of 11 December 1941.

effectively the end of his academic teaching career. In 1944, he was awarded a professorship in medical psychology at the University of Basel, but serious illness interrupted his teaching activity there soon after it began. In 1948, the C. G. Jung Institute Zurich was opened as a center for research and teaching. It was the first institute to offer formal training in Analytical Psychology. Although Jung was involved in its formation, he was no longer active as a lecturer. Even after his retirement from ETH, he maintained personal friendships with several ETH professors. Among them were H. K. Fierz, Eugen Böhler, Karl Schmid, Thadeus Reichstein, and Wolfgang Pauli. ETH honored Jung with an honorary doctorate on the occasion of his eightieth birthday in 1955.[28] In his will (1958), Jung decreed that his literary estate should be left to ETH. However, lengthy preparatory work was required before his heirs could ratify the handover of documents to ETH in 1977.[29] The corpus of writings—around 100,000 pages of manuscripts, typescript, notes etc., and around 35,000 letters—remains in the archive based in ETH library. The C. G. Jung archive has also been endowed with other Jung-related materials from several other donors. These have been open to the public since 1993 (unless legal reasons prohibit access).

Notes of the Lectures

From Jung's own surviving notes, it is evident that while he prepared meticulously for his lectures, as a seasoned speaker he gave impromptu lectures, without a written script. This means that there are no original manuscripts of his lectures. Initially his secretary, Marie-Jeanne Schmid, took notes for him. Later records made by her are unknown, however. Many of Jung's audience members may also have made notes. Mastery of shorthand was required for a continuous transcript to be made. Among the regular attendees, Rivkah Schärf, Eduard Sidler, Lucie Stutz-Meyer, and Otto Karthaus, in conjunction with Louise Tanner, had this skill and made notes of the lectures. It is not entirely clear how much Jung was aware of all these shorthand notes. Alongside ETH lectures, he gave seminars in English[30] to an international audience at

[28] Cf. footnote 7.

[29] Donation pledge by the heirs of C. G. Jung to the Swiss Confederation, March 1977.

[30] Seminars below were published between 1984 and 1997 by Princeton University Press (Bollingen Series), and Routledge London: *Analytical Psychology, given in 1925*, *Dream Analysis* (1928–1930), *The Interpretation of Visions* (1930–1934), *The Psychology of Kundalini Yoga* (1932), *Nietzsche's Zarathustra* (1934–1939).

the Psychological Club in Zurich. Those seminars were set down in minutes by a small working party and were reproduced for the use of participants. A team around Barbara Hannah, an English pupil of Jung, compiled (successive) summarized English translations[31] of Jung's ETH lectures for the same audience and in the same vein. However, these scripts made no claim to be a faithful reproduction of the text. Although German shorthand records of the original were, in part, transcribed at the time, they were not edited for publication. It is not documented whether Jung ever had any serious intention of publishing his ETH lectures. In 1959, private printings of the Hannah scripts appeared with Jung's permission through the C. G. Jung Institute, to be distributed (with restrictions) to interested parties.

The Publication of the Lectures

The publication contract with the Bollingen Foundation (BF, USA) of 1948[32] stipulated that Jung's Collected Works (*CW*) should only include texts written by the author. For this reason, the majority of the seminars and lectures did not qualify for the *CW*. In 1957, Jung agreed in principle with the Bollingen Foundation's intention also to publish his seminars.[33] It still took until 1965 before a provisional publication plan could be drawn up. It included a selection of the English-language seminars as well as a "possible selection from the ETH lectures," that is, the Hannah notes. Since the publication of the *CW* was not yet finalized, the editorial work on the seminars could not properly start until 1980. It continued to 1996.[34] In the meantime, the German original transcripts of ETH lectures had practically been forgotten. In 1993, as the end of the *Gesammelte Werke* (*GW*, German edition) loomed, the working

[31] *Notes on Lectures Given at the Eidgenössische Technische Hochschule (ETH) Zürich by Prof. Dr. C. G. Jung*, ed. Barbara Hannah, *Modern Psychology*, Vols. 1 and 2, Oct. 1933–July 1935; *Modern Psychology*, Vols. 3 and 4, Oct. 1938–March 1940; *Alchemy*, Vols. 1 and 2, Nov. 1940–July 1941.

Excerpts from the 1939–1940 lectures on psychological aspects of the *Spiritual Exercises* of St. Ignatius of Loyola were published in 1977/1978 in *Spring: A Journal for Archetypal Psychology and Jungian Thought* (1977): 183–200; (1978): 28–36. These lectures have been cited in several works of secondary literature.

[32] Publishing contract dated 15 August 1948, between C. G. Jung and the Bollingen Foundation, Washington DC.

[33] Letter from Jung to John D. Barrett of the Bollingen Foundation, dated 19 August 1957, Bollingen archive, Library of Congress.

[34] On the publication of the seminars, see *Dream Analysis Seminar*, foreword by William McGuire (Jung, 1984).

committee of the C. G. Jung heirs began to consider afresh the unpublished material. As part of this process, texts from ETH lectures gradually showed up in the C. G. Jung library in Kusnacht, in the Psychology Club in Zurich, and in ETH. In part, these were unidentified typescripts, in part shorthand records, as well as notes and diagrams. The former head of the C. G. Jung archive at ETH, Dr. B. Glaus, realized that shorthand was a dying art so he summoned two female former secretaries out of retirement and commissioned them to transcribe those notes, which could now be read only by specialists. Between 1993 and 1998, it was thus possible to bring the fragments into a coherent order by referring to old lecture programs. A series of conversations then began with Sonu Shamdasani who had just completed the editing of Jung's Kundalini seminar. An important question remained: Would the quality of the audience notes be high enough for publication? Jung had obviously not revised them. In relation to the seminars at the Psychology Club, we know he had suggested that every publication should be prefaced with a note advising that the text contained a number of errors and other imperfections.[35] This instruction should also be applied to ETH lectures. Here, there were no revised notes, other than for the Club seminars, but there were also several versions, which only partly corresponded with each other. Evidently the note takers at Jung's talks had either knowingly made only selective notes or unconsciously experienced selective perception. However, divergent versions of the same lecture often complemented each other very well. Thus, it became evident that, if possible, at least two sets of notes should be available for a reasonably reliable reconstruction. It became clear that the reconstruction of the original texts would be a hugely demanding editorial challenge, yet no project organization existed to take on this task. In 2004, the Philemon Foundation (USA) took this project on, having been founded with the purpose of bringing to completion the publication of the unpublished works of Jung. Since 2004, first Dr. Angela Graf-Nold, later Dr. Ernst Falzeder and Dr. Martin Liebscher, have been entrusted with the restoration of ETH lectures. ETH lectures are—if taken as a single corpus—Jung's most comprehensive work, and they deserve to be appropriately documented. In his written works, Jung sometimes used an academic mode of expression that some contemporary readers do not find easy to understand. However, contemporary witnesses praised Jung's qualities as *speaker*. The English-language seminars published so far were positively received,

[35] McGuire, in Jung, 1984, p. xiv.

not only because of their content, but also thanks to their accessible language. Now, in ETH lectures, an equally momentous German-language work has been made available for the first time. The list of those who have contributed to the success of this publication is long. Not all are able to witness its arrival, but the thanks of the Foundation of the Works of C. G. Jung are expressed to all.

General Introduction

ERNST FALZEDER, MARTIN LIEBSCHER,
AND SONU SHAMDASANI

BETWEEN 1933 AND 1941, C. G. Jung lectured at the Swiss Federal Institute for Technology (ETH). He was appointed a professor there in 1935. This represented a resumption of his university career after a long hiatus, as he had resigned his post as a lecturer in the medical faculty at the University of Zurich in 1914. In the intervening period, Jung's teaching activity had principally consisted in a series of seminars at the Psychology Club in Zurich, which were restricted to a membership consisting of his own students or followers. The lectures at ETH were open, and the audience for the lectures was made up of students at ETH, the general public, and Jung's followers. The attendance at each lecture was in the hundreds: Josef Lang, in a letter to Hermann Hesse, spoke of six hundred participants at the end of 1933,[36] Jung counted four hundred in October 1935.[37] Kurt Binswanger, who attended the lectures, recalled that people often could not find a seat and that the listeners "were of all ages and of all social classes: students . . . ; middle-aged people; also many older people; many ladies who were once in analysis with Jung."[38] Jung himself attributed this success to the novelty of his lectures and expected a gradual decline in numbers: "Because of the huge crowd my lectures have to be held in the *auditorium maximum*. It is of course their sensational nature that enchants people to come. As soon as people will realize that these lectures are concerned with serious matters, the numbers will become more modest."[39]

---

[36] Josef Bernhard Lang to Hermann Hesse, end of November 1933 (Hesse, 2006, p. 299).

[37] JSP, p. 87.

[38] Interview with Gene Nameche [CLM], p. 6.

[39] Jung to Jolande Jacobi, 9 January 1934 [JA].

Because of this context, the language of the lectures is far more accessible than Jung's published works at this time. Binswanger also noted that "Jung prepared each of those lectures extremely carefully. After the lectures, a part of the audience always remained to ask questions, in a totally natural and relaxed situation. It was also pleasant that Jung never appeared at the last minute, as so many other lecturers did. He, on the contrary, was already present before the lecture, sat on one of the benches in the corridor; and people could go and sit with him. He was communicative and open."[40]

The lectures usually took place on Fridays between 6 and 7 p.m. The audience consisted of regular students of technical disciplines, who were expected to attend additional courses from a subject of the humanities. But as it was possible to register as a guest auditor, many of those who had come to Zurich to study with Jung or undertake therapy attended the lectures as an introduction to Analytical Psychology. In addition, Jung also held ETH seminars with limited numbers of participants, in which he would further elaborate on the topics of the lectures. During the eight years of his lectures—which were only interrupted in 1937, when Jung travelled to India—he covered a wide range of topics. These lectures are at the center of Jung's intellectual activity in the 1930s, and furthermore provide the basis of his work in the 1940s and 1950s. Thus, they form a critical part of Jung's oeuvre, one that has yet to be accorded the attention and study that it deserves. The subjects that Jung addressed in ETH lectures are probably even more significant to present-day scholars, psychologists, psychotherapists, and the general public than they were when they were first delivered. The passing years have seen a mushrooming of interest in Eastern thought, Western hermeticism and mystical traditions, the rise of the psychological types industry and the dream work movement, and the emergence of a discipline of the history of psychology.

Contents of the Lectures

Volume 1: History of Modern Psychology *(Winter Semester 1933/1934)*

The first semester, from 20 October 1933 to 23 February 1934, consists of sixteen lectures on what Jung called the history of "modern psychology," by which he meant psychology as "a conscious science," not one that projects the psyche into the stars or alchemical processes, for

[40] Interview with Gene Nameche [CLM], p. 6.

instance. His account starts at the dawn of the age of Enlightenment, and presents a comparative study of movements in French, German, and British thought. He placed particular emphasis on the development of concepts of the unconscious in nineteenth-century German Idealism. Turning to England and France, Jung traced the emergence of the empirical tradition and of psychophysical research, and how these in turn were taken up in Germany and led to the emergence of experimental psychology. He reconstructed the rise of scientific psychology in France and in the United States. He then turned to the significance of spiritualism and psychical research in the rise of psychology, paying particular attention to the work of Justinus Kerner and Théodore Flournoy. Jung devoted five lectures to a detailed study of Kerner's work, *The Seeress of Prevorst* (1829),[41] and two lectures to a detailed study of Flournoy's *From India to the Planet Mars* (1899).[42] These works initially had a considerable impact on Jung. As well as elucidating their historical significance, his consideration of them enables us to understand the role that his reading of them played in his early work. Unusually, in this section Jung eschewed a conventional history of ideas approach, and placed special emphasis on the role of patients and subjects in the constitution of psychology. In the course of his reading of these works, Jung developed a detailed taxonomy of the scope of human consciousness, which he presented in a series of diagrams. He then presented a further series of illustrative case studies of historical individuals in terms of this model: Niklaus von der Flüe, Goethe, Nietzsche, Freud, John D. Rockefeller, and the "so-called normal man."

Of the major figures in twentieth-century psychology, Jung was arguably the most historically and philosophically minded. These lectures thus have a twofold significance. On the one hand, they present a seminal contribution to the history of psychology, and hence to the current historiography of psychology. On the other hand, it is clear that the developments that Jung reconstructed teleologically culminate in his own "complex psychology" (his preferred designation for his work), and thus present his own understanding of its emergence. This account provides a critical correction to the prevailing Freudocentric accounts of the development of Jung's work, which were already in circulation at this time. The detailed taxonomy of consciousness that he presented in the second part of this semester was not documented

---

[41] Kerner (1829).

[42] Flournoy (1900 [1899]).

in any of his published works. In presenting it, Jung noted that the difficulties which he had encountered with his project for a psychological typology had led him to undertake this. Thus these lectures present critical aspects of Jung's mature thought that are unavailable elsewhere.

Volume 2: **The Psychology of Consciousness and Dream Psychology** *(Summer Semester 1934)*

This volume presents twelve lectures from 20 April 1934 to 13 July 1934. Jung commenced with lectures on the problematic status of psychology, and attempted to give an account as to how the various views of psychology in its history, which he had presented in the first semester, had been generated. This led him to account for national differences in ideas and outlook, and to reflect on different characteristics and difficulties of the English, French, and German languages when it came to expressing psychological materials. Reflecting on the significance of linguistic ambiguity led Jung to give an account of the status of the concept of the unconscious, which he illustrated with several cases. Following these general reflections, he presented his conception of the psychological functions and types, illustrated by practical examples of their interaction. He then gave an account of his concept of the collective unconscious. Filling a lacuna in his earlier accounts, he gave a detailed map of the differentiation and stratification of its contents, in particular as regards cultural and "racial" differences. Jung then turned to describing methods for rendering accessible the contents of the unconscious: the association experiment, the psycho-galvanic method, and dream analysis. In his account of these methods, Jung revised his previous work in the light of his present understanding. In particular, he gave a detailed account of how the study of associations in families enabled the psychic structure of families and the functioning of the complexes to be studied. The semester concluded with an overview of the topic of dreams and the study of several dreams.

On the basis of his reconstruction of the history of psychology, Jung then devoted the rest of this and the following semesters to an account of his "complex psychology." As in the other semesters, Jung was confronted with a general audience, a context that gave him a unique opportunity to present a full and generally accessible account of his work, as he could not presuppose prior knowledge of psychology. Thus we find here the most detailed, and perhaps most accessible, introduction to his own theory. This is by no means just an introduction to previous work, however, but a full-scale reworking of his early work in terms of his current understanding, and it presents models of the personality that cannot be found anywhere else in his work. Thus, this volume is Jung's most up-to-date account of his theory of complexes, association experiments, understanding of dreams, the structure of the personality, and the nature of psychology.

Volume 3: Modern Psychology and Dreams *(Winter Semester 1934/1935 and Summer Semester 1935)*

The third volume presents lectures from two consecutive semesters: seventeen lectures from 26 October 1934 to 8 March 1935, and eleven lectures from 3 May 1935 to 12 July 1935, here collected in one volume as they all deal primarily with possible methods to access, and try to determine the content of, the unconscious. Jung starts with a detailed description of Freud's and, to a somewhat lesser extent, Adler's theory and method of analyzing dreams, and then proceeds to his own views (dreams are "pure nature" and of a complementary/compensatory character) and technique (context, amplification). He focuses particularly on three short dream series, the first from the Nobel Prize winner Wolfgang Pauli, the second from a young homosexual man, and the third from a psychotic person, using them to describe and interpret special symbolisms. In the following semester, he concludes the discussion of the mechanism, function, and use of dreams as a method to enlighten us and to get to know the unconscious, and then draws attention to "Eastern parallels," such as yoga, while warning against their indiscriminate use by Westerners. Instead he devotes the rest of the semester to a detailed example of "active imagination," or "active phantasizing," as he calls it here, with the help of the case of a fifty-five-year-old American lady, the same case that he discussed at length in the German seminar of 1931.

This volume gives a detailed account of Jung's understanding of Freud's and Adler's dream theories, shedding interesting light on the points in which he concurred and in which he differed, and how he developed his own theory and method in contradistinction to those. Since he was dealing with a general audience, a fact that he was very much aware of, he tried to stay on a level as basic as possible—which is also of great help to the contemporaneous, nonspecialized reader. This is also true for his method of active imagination, as exemplified in one long example. Although he used material also presented elsewhere, the present account is highly interesting precisely because it is tailored to a most varied general audience, and differs accordingly from presentations given to the hand-picked participants in his "private" seminars, or in specialized books.

Volume 4: Psychological Typology *(Winter Semester 1935/1936 and Summer Semester 1936)*

The fourth volume also combines lectures from two semesters: fifteen lectures from 25 October 1935 to 6 March 1936, and thirteen lectures from 1 May 1936 to 10 July 1936. The winter semester gives a general introduction into the history of typologies, and typology in intellectual and religious history, from antiquity to Gnosticism and Christianity, from Chinese philosophy (yin/yang) to Persian religion and philosophy (Ahriman/Lucifer), from the French revolution ("déesse raison") to Schiller's *Letters on the Aesthetic Education of Man.* Jung introduces and describes in detail the two attitudes (introversion and extraversion) and the four functions (thinking and feeling as rational functions, sensation and intuition as irrational functions). In the summer semester, he focuses on the interplay between the attitudes and the various functions, detailing the possible combinations (extraverted and introverted feeling, thinking, sensation, and intuition) with the help of many examples.

This volume offers an excellent, first-hand introduction to Jung's typology, and is *the* alternative for contemporaneous readers who are looking for a basic, while authentic text, as opposed to Jung's magnum opus *Psychological Types*, which, as it were, hides the sleeping beauty behind a thick wall of thorny bushes, namely, its 400 plus pages of "introduction," only after which Jung deals with his own typology proper. As in the previous volumes, readers will benefit from the fact that Jung was compelled to give a basic introduction to and overview of his views.

Volume 5: Psychology of the Unconscious *(Summer Semester 1937 and Summer Semester 1938)*

Jung dedicated his lectures of summer 1937 (23 April–9 July; eleven lectures) and summer 1938 (29 April–8 July; ten lectures) to the psychology of the unconscious. The understanding of the sociological and historical dependency of the psyche and the relativity of consciousness form the basis to familiarize the audience with different manifestations of the unconscious related to hypnotic states and cryptomnesia, unconscious affects and motivation, memory and forgetting. Jung shows the normal and pathological forms of invasions of unconscious contents into consciousness and outlines the methodologies to bring unconscious material to the surface. This includes methods such as the association experiment, dream analysis, active imagination, as well as different forms of creative expression, but also ancient tools of divination including astrology and the I-Ching. The summer semester of 1938 returned to the dream series of the young homosexual man discussed in detail in the lectures of 1935, this time highlighting Jung's method of dream interpretation on an individual and a symbolic level.

Jung illustrates his lectures with several diagrams and clinical cases to make it more accessible to nonpsychologists. In some instances the lectures provide welcome additional information to published articles, as Jung was not obliged to restrict his material to a confined space. For example, Jung elaborated on the famous case of the so-called moon-patient, which was so important for his understanding of psychic reality and psychosis, or gave a very personal introduction to the usage of the I-Ching. The lectures also shed a new historical light on his journeys to Africa, India, and New Mexico and his reception of psychology, philosophy, and literature.

Volume 6: **Psychology of Yoga and Meditation** *(Winter Semester 1938/1939 and Summer Semester 1939; plus the First Two Lectures of the Winter Semester 1940/1941)*

The lecture series of the winter semester 1938/1939 (28 October–3 March; fifteen lectures) and the first half of the summer semester 1939 (28 April–9 June; six lectures) are concerned with Eastern spirituality. Starting out with the psychological concept of active imagination, Jung seeks to find parallels in Eastern meditative practices. His focus is directed on meditation as taught by different yogic traditions and in Buddhist practice. The texts for Jung's interpretation are Patanjali's *Yoga Sûtra*, according to the latest research written around 400 CE[43] and regarded as one of the most important sources for our knowledge of yoga today, the *Amitâyur-dhyâna-sûtra* from the Chinese Pure Land Buddhist tradition, translated from Sanskrit to Chinese by Kâlayasas in 424 CE,[44] and the *Shrî-Chakra-Sambhâra Tantra*, a scripture related to tantric yoga, translated and published in English by Arthur Avalon (Sir John Woodroffe) in 1919.[45]

---

[43] Maas (2006).

[44] Müller (1894), pp. xx–xxi.

[45] Avalon (1919).

Nowhere else in Jung's works can one find such detailed psychological interpretations of those three spiritual texts. In their importance for understanding Jung's take on Eastern mysticism, the lectures of 1938/39 can only be compared to his reading of the *Secret of the Golden Flower*[46] or the seminars on Kundalini Yoga.[47]

In the winter semester 1940/41, Jung summarizes the arguments of his lectures on Eastern meditation. The summary is published as an addendum at the end of this volume.

Volume 7:* Spiritual Exercises of Ignatius of Loyola *(Summer Semester 1939 and Winter Semester 1939/1940; in Addition: Lecture 3, Winter Semester 1940/1941)

The second half of the summer semester 1939 (16 June–7 July; four lectures) and the winter semester 1939/40 (3 October–8 March; sixteen lectures) were dedicated to the *Exercitia Spiritualia*[48] of Ignatius of Loyala, the founder and first general superior of the Society of Jesus (Jesuits). As a knight and soldier, Ignatius was injured in the battle of Pamplona (1521), in the aftermath of which he experienced a spiritual conversion. Subsequently he renounced his worldly life and devoted himself to the service of God. In March 1522, the Virgin Mary and the infant Jesus appeared to him at the shrine of Montserrat, which led him to search for solitude in a cave near Manresa. There he prayed for seven hours a day and wrote down his experiences for others to follow. This collection of prayers, meditations, and mental exercises built the foundation of the *Exercitia Spiritualia* (1522–1524). In the text, Jung saw the equivalent to the meditative practice of the Eastern spiritual tradition. He provides a psychological reading of it, comparing it to the modern Jesuit understanding of theologians like Erich Przywara.

Jung's considerations on the *Exercitia Spiritualia* follow the lectures on Eastern meditation of the previous year. Nowhere in Jung's writings is there to be found a similarly intense comparison between oriental and occidental spiritualism. Its approach equals the aim of the annual Eranos conference, namely to open up a dialogue between the East and the West. Jung's critical remarks about the embrace of Eastern mysticism by modern Europeans and his suggestion to the latter to come back to their own traditions are illuminated through those lectures.

---

[46] Jung (1929).
[47] Jung (1932).
[48] Ignatius of Loyala (1996 [1522–1524]).

In the winter semester 1940/1941, Jung dedicated the third lecture to a summary of his lectures on the *Exercitia Spiritualia*. This summary is added as an addendum to volume 7.

Volume 8: The Psychology of Alchemy *(Winter Semester 1940/1941 and Summer Semester 1941)*

The lectures of the winter semester 1940/41 (from lecture 4 onward; 29 November–28 February; twelve lectures) and the summer semester 1941 (2 May–11 July; eleven lectures) provide an introduction to Jung's psychological understanding of alchemy. He explained the theory of alchemy, outlined the basic concepts, and gave an account of psychological research into alchemy. He showed the relevance of alchemy for the understanding of the psychological process of individuation. The alchemical texts that Jung talked about included, next to famous examples such as the *Tabula Smaragdina* and the *Rosarium Philosophorum*, many less well-known alchemical treatises.

The lectures on alchemy built a cornerstone in the development of Jung's psychological theory. His Eranos lectures from 1935 and 1936 were dedicated to the psychological meaning of alchemy and were later merged together in *Psychology and Alchemy* (1944). The ETH lectures on alchemy highlight the way Jung's thinking of alchemy developed through those years. As an introduction to alchemy, they provide an indispensable tool in order to understand the complexity of his late works such as *Mysterium Coniunctionis*.

Editorial Guidelines

With the exception of a few preparatory notes, there is no written text by Jung. The present text has been reconstructed by the editors through several notes by participants of Jung's lectures. Through the use of short-hand, the notes taken by Eduard Sidler, a Swiss engineer, and Rivkah Schärf—who later became a well known religious scholar, psychotherapist, and collaborator of Jung—provide a fairly accurate first basis for the compilation of the lectures. (The short-hand method used is outdated and had to be transcribed by experts in the field.)

Together with the recently discovered scripts by Otto Karthaus, who made a career as one of the first scientific vocational counselors in Switzerland, Bertha Bleuler, and Lucie Stutz-Meyer, the gymnastic teacher of the Jung family, these notes enable us to not only regain access to the contents of Jung's orally delivered lectures, but also to get a feeling for the fascination of the audience with Jung the orator.

There also exists a set of mimeographed notes in English that have been privately published and circulated in limited numbers. They were edited and translated by an English-speaking group in Zurich around Barbara Hannah and Elizabeth Welsh, and present more of a résumé than an attempt at a verbatim account of the content of the lectures. For the first years Hannah's edition relied only on the notes by Marie-Jeanne Schmid, Jung's secretary at the time; for the later lectures the script of Rivkah Schärf provided the only source for most of the text. The edition was disseminated in private imprints from 1938 to 1968.

The Hannah edition does deviate from Jung's original spoken text as recorded in the other notes. Hannah and Welsh stated in their "Prefatory Note" that their compilation did "not claim to be a verbatim report or literal translation." Hannah was mainly interested in the creation of a readable and consistent text and did not shy away from adding or omitting passages for that purpose. As her edition was only based on one set of notes she could not correct passages where Schmid or Schärf rendered Jung's text wrongly. But as Hannah had the advantage of talking to Jung in person, when she was not sure about the content of a certain passage, her English compilation is sometimes useful to provide additional information to the readers of our edition.

In contrast to a critical edition, it is not intended to provide the differing variations in a separate critical apparatus. Had we faithfully listed all the minor or major variants in the scripts, the text would have become virtually unreadable and thus would have lost the accessibility that is the hallmark of Jung's presentation. For the most part, however, we can be reasonably certain that the compilation accurately reflects what Jung said, although he may have used different words or formulations. Moreover, in quite a number of key passages it was even possible to reconstruct the verbatim content, for example, when different note takers identified certain passages as direct quotes. Variations often do not add to the content and intelligibility, and often originated in errors or lack of understanding by the participant taking notes. In their compilation, the editors have worked according to the principle that as much information as possible should be extracted from the manuscripts. If there are obvious contradictions that cannot be decided by the editor, or, as might be the case, clear errors on behalf of Jung or the listener, it will be clarified by the editor's annotation.

Of the note takers, Eduard Sidler, whose background was in engineering, had the least understanding of Jungian psychology at the beginning, although naturally he became more familiar with Jungian psychology over time. In any case, he did try to protocol faithfully as much as he could, making his the most detailed notes. Sometimes he could no longer follow, however, or clearly misunderstood what was said. On the other hand, we have Welsh and Hannah's version, which in itself was already a collation and obviously heavily edited, but is (at least for the first semesters) the most consistent manuscript and also contains things that are missing in other notes. Moreover, they state that "Prof. Jung himself . . . has been kind enough to help us with certain passages," although we do not know which these are. In addition, over the course of the years, and also for individual lectures, the quality, accuracy, and reliability of the scripts by the different note takers vary, as is only natural. In short, the best we can do is try and find an approximation of what Jung actually said. In essence, it will always have to be a judgment call how to collate those notes.

It is thus impossible to establish exact editorial principles for each and every situation, so that different editors would inevitably arrive at exactly the same formulations. We could only adhere to some general guidelines, such as "Interfere as little as possible, and as much as necessary," or "Try to establish what the most likely thing was that Jung might have said, on the basis of all the sources available" (including the *Collected Works*, autobiographical works or interviews, other seminars, interviews, etc.). If two transcripts concur, and the third is different, it is usually safe to go with the first two. In some cases, however, it is clear from the context that the two are wrong, and the third is correct. Or if all three of them are unclear, it is sometimes possible to "clean up" the text by having recourse to the literature, for instance, when Jung summarizes Kerner's story of the Seeress of Prevorst. As with all scholarly works of this kind, there is no explicit recipe that can be fully spelled out: One has to rely on one's scholarly judgement.

These difficulties not only concern the establishment of the text of Jung's ETH lectures, but also pertain to notes of his seminars in general, many of which have already appeared in print without addressing this problem. For instance, the introduction to the *Dream Analysis* seminar mentions the number of people that were involved in preparing the notes, but there is no account of how they worked, or how they established the text (Jung, 1984, pp. x–xi). Some manuscript notes in the library of the Analytical Psychology Club in Los Angeles indicate that the compilation of the notes involved significant "processing by committee." It is interesting in this regard to compare the sentence structure of the *Dream Analysis* seminar with the 1925 seminar, which was checked by Jung. On 19 October 1925, Jung wrote to Cary Baynes, after checking her notes and acknowledging her literary input: "I faithfully worked through the notes as you will see. I think they are as a whole very accurate. Certain lectures are even fluent, namely those which you could not stop your libido from flowing in" (Cary Baynes papers, contemporary medical archives, Wellcome Library, London).

Our specific situation seems to be a "luxury" problem, as it were, because we have several transcripts, which was often not the case in other seminars. We also have the disadvantage of no longer being able to ask Jung himself, as for instance Cary Baynes, Barbara Hannah, Marie-Jeanne Schmid, or Mary Foote could do. We can only work as best we can, and caution the reader that there is no guarantee that this is "verbatim Jung," although we have tried to come as close as possible to what he actually said.

Abbreviations

CLM Jung biographical archive, Countway Library of Medicine, Boston.

JA Jung collection, History of Science Collection: Swiss Federal Institute of Technology Archive, Zurich.

JSP McGuire, William & R. F. C. Hull (eds.) (1977). *C. G. Jung Speaking. Interviews and Encounters. Princeton: Princeton University Press.*

References

Avalon, Arthur [Sir John Woodroffe] (ed.) (1919). *Shrî-chakra-sambhâra Tantra*. Trans. Kazi Dawa-Samdup. *Tantrik Texts*, vol. 7. London: Luzac & Co; Calcutta: Thacker, Spink & Co.

Flournoy, Théodore (1900 [1899]). *Des Indes à la planète Mars. Étude sur un cas de somnambulisme avec glossolalie*. Paris, Geneva: F. Alcan, Ch. Eggimann. *From India to the Planet Mars. A Case of Multiple Personality with Imaginary Languages*. With a Foreword by C. G. Jung and Commentary by Mireille Cifali. Ed. and introduced by Sonu Shamdasani. Princeton: Princeton University Press, 1994.

Hesse, Hermann (2006 [1916–1944]). *"Die dunkle und wilde Seite der Seele": Briefwechsel mit seinem Psychoanalytiker Josef Bernhard Lang 1916–1944*. Ed. Thomas Feitknecht. Frankfurt am Main: Suhrkamp.

(Saint) Ignatius of Loyola (1996 [1522–1524]). *The Spiritual Exercises*, in *Personal Writings: Reminiscences, Spiritual Diary, Selected Letters Including the Text of The Spiritual Exercises*. Trans. with introductions and notes by Joseph A. Munitiz and Philip Endean. London: Penguin, pp. 281–328.

Jung, C. G. (1929). Commentary on "The Secret of the Golden Flower." *CW* 13.

Jung, C. G. (1932). *The Psychology of Kundalini Yoga: Notes of the Seminar Given in 1932 by C. G. Jung*. Ed. Sonu Shamdasani. Bollingen Series XCIX. Princeton: Princeton University Press, 1996.

Kerner, Justinus Andreas Christian (1829). *Die Seherin von Prevorst. Eröffnungen über das innere Leben und über das Hineinragen einer Geisterwelt in die unsere*. Two vols. Stuttgart, Tubingen: J. G. Cotta'sche Buchhandlung. 4., vermehrte und verbesserte Auflage: Stuttgart, Tubingen: J. G. Cotta'scher Verlag. Reprint: Kiel: J. F. Steinkopf Verlag, 2012. *The Seeress of Prevorst, Being Revelations Concerning the Inner-Life of Man, and the Inter-Diffusion of a World of Spirits in the One We Inhabit*. Trans. Catherine Crowe. London: J. C. Moore, 1845. Digital reprint: Cambridge: Cambridge University Press, 2011.

Maas, Philipp A. (2006). *Samâdhipâda: das erste Kapitel des Pâtañjalayogaśâstra zum ersten Mal kritisch ediert*. Aachen: Shaker.

Müller, Max (1894). *Introduction to Buddhist Mahâyâna Texts, The Sacred Books of The East*, vol. 49. Ed. Max Müller. Oxford: The Clarendon Press.

Introduction to Volume 1

ERNST FALZEDER

This volume presents the collation of various lecture notes that were taken of the sixteen lectures that Jung gave during the first semester of his professorship at the *Eidgenössische Technische Hochschule* in Zurich, from 20 October 1933 to 23 February 1934. He entitled them "modern psychology," thus partly realizing a program that he had described as a task of the future some three years earlier. There is "a particular current of thought," he had written,

> which can be traced back to the Reformation. Gradually it freed itself from innumerable veils and disguises, and it is now turning into the kind of psychology which Nietzsche foresaw with prophetic insight—the discovery of the psyche as a new fact. Some day we shall be able to see by what tortuous paths modern psychology has made its way from the dingy laboratories of the alchemists, via mesmerism and magnetism (Kerner, Ennemoser, Eschenmayer, Passavant, and others), to the philosophical anticipations of Schopenhauer, Carus, and von Hartmann; and how, from the native soil of everyday experience in Liébeault and, still earlier, in Quimby (the spiritual father of Christian Science), it finally reached Freud through the teachings of the French hypnotists. This current of ideas flowed together from many obscure sources, gaining rapidly in strength in the nineteenth century and winning many adherents, amongst whom Freud is not an isolated figure (1930a, § 748).

His account in the lectures starts at the dawn of the age of the Enlightenment, and presents a comparative study of movements in French, German, and British thought. He placed particular emphasis on the development of conceptions of the unconscious in nineteenth-century German Idealism. Turning to England and France, Jung traced the emergence of the empirical tradition and psychophysical research, and how these in turn became taken up in Germany and led to the emergence of experimental psychology. He reconstructed the rise of scientific psychology in France and the United States. In essence he described this as a constant development from a naïve "psychology," which found the psychical contents where it had unconsciously projected them beforehand into the outer world (as in astrology, for instance), to "modern" psychology, that is, to psychology "as a conscious science," as he puts it in these lectures.

As he wrote elsewhere,

> the projections falling back into the human soul caused such a terrific activation of the unconscious that in modern times man was compelled to postulate the existence of an unconscious psyche. The first beginnings of this can be seen in Leibniz and Kant, and then, with mounting intensity, in Schelling, Carus, and von Hartmann, until finally modern psychology discarded the last metaphysical claims of the philosopher-psychologists and restricted the idea of the psyche's existence to the psychological statement, in other words, to its phenomenology (1941, § 375).

In the second part of the lectures, he then turned to the significance of spiritualism and psychical research in the rise of psychology, giving particular attention to the work of Justinus Kerner and Théodore Flournoy. Jung devoted five lectures alone to a detailed study of Kerner's work, *The Seeress of Prevorst* (1829), and two more lectures to one of Flournoy's, *From India to the Planet Mars* (1899). These works initially had a considerable impact on Jung that can hardly be overestimated. As well as elucidating their historical significance, his consideration of them enables us to understand the role that his reading of them played in his early work. Unusually, in this section, Jung eschewed a conventional history of ideas approach, and placed special emphasis on the role of patients and subjects in the constitution of psychology. In the course of his reading of these works, Jung also developed a detailed taxonomy of the scope of human consciousness, which he presented in a series of diagrams. Finally, he presented a further series of illustrative case studies of historical individuals in terms of this model: Niklaus von der Flüe, Goethe, Nietzsche, Freud, John D. Rockefeller, and "so-called normal man."

Of the major figures in twentieth-century psychology, Jung was arguably the most historically and philosophically minded. These lectures thus have a twofold significance: On the one hand, they present a seminal contribution to the history of psychology, and hence to the contemporary historiography of psychology in general. On the other hand, it is equally clear that the developments which Jung reconstructed teleologically culminate, in his account of them, in his own "complex psychology" (his preferred designation for his work),[49] and thus present his own understanding of its emergence.

Jung was by no means the first to give a history of the budding science of psychology, however. Let us mention, in chronological order of their appearance, only the works by Théodule Ribot (1870, 1879), Eduard von Hartmann (1901), Max Dessoir (1902, 1911), G. Stanley Hall (1912), James Mark Baldwin (1913), Pierre Janet (1919), and Edwin Boring (1929). With regard to the development of so-called depth psychology, there is also Freud's account in *On the History of the Psycho-Analytic Movement* (1914). Practically all distinguish, like Jung, between a prescientific and scientific period of the discipline, which some of them—such as Baldwin, Dessoir, Hall, and von Hartmann—also expressly call "modern psychology," although they differ as to when the latter began. Von Hartmann (1901, p. 1) dates the beginning of "modern psychology" to the middle of the nineteenth century, for instance. Dessoir (1911 [1912], pp. 221 ff.) states that modern French psychology begins with Condillac, while Baldwin (1913, p. 95) maintains that "[w]ith the development of the dualism between mind and body up to the stage it reached in René Descartes . . . , the period properly to be called 'modern' commences."

Many of these authors stress that psychology should be modeled after the natural sciences. In the early account of Ribot (1870, p. 19), for example, "experimental psychology alone makes up the whole of psychology, the rest being matters of philosophy or metaphysics, and therefore outside of science." In 1879 (p. ii), he speaks of "cette séparation, qui devient chaque jour plus nette, entre l'ancienne et la nouvelle psychologie" [this separation between old and new psychology, which becomes clearer every day] and declares: "l'ancienne psychologie est condamné" [the old psychology is overthrown]. Fifty-nine years later, Boring (1929) concentrates only on experimental psychology in the first place, and tries to show how physics, as an experimental science, had set the pace for physiology, and physiology proceeded to do the same for psychology. Thus, Newton and Young led on to Purkinje, Weber, and Johannes Müller, to Fechner and Helmholtz, and so to Wundt and the psychological laboratory. (Boring also

[49] In his preface to Toni Wolff's *Studies in Jungian Psychology*, he credited her with introducing this designation (Jung, 1959, § 887).

stressed that to become sophisticated the experimental psychologist needs a historical perspective, particularly in systematic or theoretical questions. Despite his theoretical orientation, by the way, he did seek personal help in psychoanalytic treatment for a year in 1933, with the Freudian Hanns Sachs in Boston, doing five sessions a week, but both agreed later that it did not help much, if at all.)

As in Boring's account, the dominant narrative that emerged was a view that developments culminated in an ever more natural scientific methodology, based on physiology, reaching a high point in Wundt's experimental program.

There are, however, some prominent exceptions to this conceptualization of psychology and its history. Eduard von Hartmann, although maintaining that "modern psychology" should be oriented towards the natural sciences (1901, p. 1), holds that it became, mainly through his own work, the "science of the unconscious": "Just as natural science deals with conscious-less matter, psychology deals with unconscious psychic material" (ibid., p. 13). "Pure psychology of consciousness is impossible," and "psychology is that science that investigates how conscious psychic phenomena depend, according to the laws of nature, on what lies beyond consciousness" (ibid., p. 25). He acknowledges, however, that leading protagonists such as Fechner, Lotze, and Wundt (and later Brentano) contested the existence of the unconscious, so that the focus of modern psychology shifted to the explanation of psychic matters through physiological dispositions and processes (ibid., p. 14).

G. Stanley Hall's account (1912) may be of special interest in this context, because he personally knew both Jung and Freud, and was their host when they came to Worcester, Massachusetts, to deliver their lectures at Clark University in 1909. He gives Hartmann a prominent place in his studies of six founders of modern psychology, besides Zeller, Lotze, Fechner, Helmholtz, and Wundt (with four of whom he had personally studied, except Zeller and Lotze). "[M]en like Hartmann," he wrote, "are true representatives of the modern spirit, for he conserves rather than ignores the best in the past" (ibid., p. 238). So, while holding fast to the importance of the natural sciences and their results, by the "erection of the Unconscious as a world principle" (ibid., p. 238), he was "at least partially satisfying the metaphysical needs of his contemporaries" and "made philosophy again an enthusiasm" (ibid., p. 191). When Hall points out that "Hartmann's chief significance lies in his advocacy of the Unconscious and his opposition to the 'consciousness philosophies'" (ibid., p. 239), we may understand why Jung counted him, along with

Kant, Schopenhauer, and C. G. Carus, among those who "had provided him with the tools of thought" (McGuire & Hull, 1977, p. 207).

Janet's book (1919) is essentially a thorough and systematic overview of the contemporary forms of psychotherapy and their history, beginning with the early magnetists, whom he had rediscovered, and hypnotism, and is naturally tinged by his own theories of the "subconscious" and "psychological analysis," as he called it. This is not the place to enter into the old question of priority between Janet and Freud, but there is no doubt that many of Janet's earlier ideas bear great resemblance to the views of Breuer and Freud, and he created a huge oeuvre that waits to be rediscovered. Jung repeatedly mentioned his debt to Janet, with whom he had studied in the winter semester of 1902/1903, and stated explicitly: "I do not come from Freud, but from Eugen Bleuler and Pierre Janet, who were my direct teachers" (1934 [1968], after § 1034; my trans.). Henri Ellenberger, who was inspired by Janet's line of "dynamic psychology," ends his long chapter on him thus: "Janet's work can be compared to a vast city buried beneath ashes, like Pompeii. The fact of any buried city is uncertain. It may remain buried forever. It may remain concealed while being plundered by marauders. But it may also perhaps be unearthed some day and brought back to life." At the time, however, "while the veil of Lesmosyne was falling upon Janet, the veil of Mnemosyne was lifted to illuminate his great rival, Sigmund Freud" (1970, p. 409).

This great rival wrote his own, highly subjective view of the *History of the Psycho-Analytic Movement* (1914), Freud's only openly polemical work, which set the tone for the historiography of psychoanalysis and analytical psychology for a long time, and lay the basis for the Freudo-centric reading of the origins of analytical psychology. Its secret purpose was, as becomes abundantly clear from Freud's correspondence with the members of the so-called Secret Committee, to get rid of Jung and the Zurichers. Accordingly, Jung's emerging different theory and methodology were heavily criticized as being "obscure, unintelligible and confused," suggesting that much of this "lack of clearness" is due "to lack of sincerity" (ibid., p. 60).

Jung's work on the associations experiment and on the psychology of dementia praecox were both annexed to psychoanalysis. For Freud, these only consisted in the application of the theory and procedures of psychoanalysis into areas that it had as yet not been utilized in—experimental psychology and psychiatry—for the simple reason that prior to the interest taken in psychoanalysis at the Burghölzli no other major psychiatric hospital or university clinic had extensively permitted such research. Freud's assessment of Jung's work in his history was that what was valuable in it lay in its application and extension of his own discoveries, whereas Jung's supposedly new innovations would represent a secession. Jung did not publically reply to Freud's account at the time. However, in critical respects, the comprehensive intellectual history that Jung presents in these lectures can be taken in part to constitute an attempt at a reply and a rebuttal.

Significantly, all of these authors were psychologists themselves (with the possible exception of Freud, who was a trained neurologist), who in their own work claimed to be establishing the one true "scientific" psychology. Writing the history of psychology had become one means towards this end, through constructing the genealogical lineages that culminated in their respective own work, and discrediting rival claimants to the throne. In this regard, Jung's own account follows this pattern. For many decades, before the history of psychology started to become a properly historical discipline, it was largely Edwin Boring's history that held sway in the field of psychology.

Against this background, it becomes clear how Jung's own presentation of the history of "modern" psychology is an effort to situate himself and his theory in this tradition—be it by distancing himself from certain trends, be it by presenting himself as someone who carried out and further developed others, and, finally—in the coming semesters—as someone in whose new findings this development culminated.

At the outset, he made it clear that he would "attempt to convey . . . a sense of the field known as 'psychology,'" in charting a path through the "incredible chaos of opinions" that characterized that field. He deliberately chose a quite general title—"modern psychology"—as he went on saying, "because the matters at hand are of a very general nature. Instead of engaging with specific doctrines, my aim is to paint a picture based on immediate experience in order to depict the development of modern psychological ideas." When he twice received reactions from students that the lectures did not meet their expectations and, specifically, that the topics and case histories would be too far-fetched and historical in their opinion, and when they wished that Jung would talk more about contemporary problems and his own psychological theory, he stressed again: "You must bear in mind that I set out to give a course of lectures on modern psychology, and I cannot claim that modern psychology is identical with myself. It would be most immodest if I advanced my own views and opinions more than I already have."

On the other hand, it is quite obvious that his presentation of the history of "modern" psychology gave a highly selective account of various philosophical systems, focusing on what they had to say about *unconscious* aspects of the human psyche, or "the" unconscious in general, and particularly stressing those characteristics of unconscious motives and contents that would play a crucial part in his own theory, for instance, the autonomy of the soul/psyche and its contents (e.g., complexes), the "objectivation" of unconscious contents, or the importance of "primordial images." In fact, these lectures could have also be entitled, "A History of the Unconscious," leading up to and culminating in Jung's own concepts—or even, to put it provocatively, "My Predecessors."

Talking about Kant, for example, he hardly dealt with the latter's theory of cognition (apart from mentioning Kant's relativization of the concepts of space and time), which had a tremendous impact and brought about a "Copernican Turn" in philosophy, but rather with his *Dreams of a Spirit-Seer* and Kant's opinion "that the human soul . . . forms an indissoluble communion with all immaterial natures of the spirit-world," or his concept of the field of "obscure representations" being "the largest in the human being." With regard to Schopenhauer's work, he stressed above all that he was "the first to declare that the human psyche means suffering," and that he "might as well have referred to [the] 'will' as the 'unconscious,'" but neglected his Kantian heritage and epistemological works, as well as his teaching of how to escape the constant suffering, namely, through negation of the will and asceticism. He also passed over Schopenhauer's formulations of what Freud would later term defense mechanisms, or his critique of the *principium individuationis*, which stands in stark contrast to Jung's own notion of individuation. In the case of Nietzsche, Jung did not enter into a discussion of his philosophy—at least not in the lectures of this semester—but cited a passage of his as an example of cryptomnesia, and used him as a case history with regard to his diagram of the fields of consciousness.

Here a word might be in order on Jung's attitude towards philosophy and epistemology in general. Although he credited "[c]ritical philosophy" with being "the mother of modern psychology" (1954 [1939], § 759),[50] he emphasized time and again, throughout his writings, lectures, seminars, and interviews, that he was not a philosopher,

[50] "The development of Western philosophy during the last two centuries has succeeded in isolating the mind in its own sphere and in severing it from its primordial oneness with the universe" (ibid.).

but an empiricist. Here are a few choice quotations: "Not being a philosopher, but an empiricist . . ." (1926, § 604). "I am an empiricist, not a philosopher" (1938 [1954], § 149). "Although I have often been called a philosopher, I am an empiricist" (1939 [1937], § 2). "I define myself as an empiricist" (1962, only in German edition, p. 375). "You see, I am not a philosopher. I am not a sociologist—I am a medical man. I deal with facts. This cannot be emphasized too much" (*Jung Speaking*, 1977, p. 206). "I am an empiricist, with no metaphysical views at all" (ibid., p. 414). "You criticize me as though I were a philosopher. But you know very well that I am an empiricist" (1975, letter of 25 April 1955, p. 246). "[M]y concepts are based on empirical findings . . . I speak of the facts of the living psyche and have no use for philosophical acrobatics" (1945b, § 438). "My business is merely the natural science of the psyche, and my main concern to establish the facts" (1946a, § 537). In other words, he claimed to have freed psychology from the acrobatics and "phantasmagoric speculations of philosophers" (1955/1956, § 53), that is, from unconscious projections that posed as philosophical insights, and turned it into an empirical science. In short: "Everyone who says I am a mystic is just an idiot"! (*Jung Speaking*, 1977, p. 333)

But things are not that simple. In quite a number of instances, we find also very positive statements about philosophy. At times he even admitted that he himself was a philosopher at heart: "I have always been of the opinion that Hegel is a psychologist in disguise, just as *I am a philosopher in disguise*" (1973, p. 194; my ital.). In analytical practice, too, "we psychotherapists ought really to be philosophers or philosophic doctors—or rather . . . we already are so, though we are unwilling to admit it" (1943 [1942], § 181). Analysis "is something like antique philosophy" (*Jung Speaking*, p. 255). In fact, he criticized Freud for being *too much* of an empiricist: Freud "proceeded quite empirically" (1934a, § 212), but it would have been "a great mistake on Freud's part to turn his back on philosophy," whereas he, Jung, had "never refused the bitter-sweet drink of philosophical criticism," which "has helped me to see that every psychology—my own included—has the character of a subjective confession" (1950 [1929], § 774). How does this jibe with Jung's emphatic statement that his readers "should never forget . . . that I am not making a confession" (1951, p. x)?

In one instance Jung even admits that a purely empirical approach is impossible in psychology because psychology, just like philosophy, is a system "of opinion about *objects which cannot be fully experienced and therefore cannot be adequately comprehended by a purely empirical approach. . . . Neither discipline can do without the other*" (1931a, § 659; my ital.).

Sometimes Jung voiced doubts about the overall validity of his conclusions:

> I fancied I was working along the best scientific lines, establishing facts, observing, classifying, describing causal and functional relations, only to discover in the end that I had involved myself in a net of reflections which extend far beyond natural science and ramify into the fields of philosophy, theology, comparative religion, and the humane sciences in general. This transgression, as inevitable as it was suspect, has caused me no little worry. . . . [I]t seemed to me that my reflections were suspect also in principle. . . . There is no medium for psychology to reflect itself in: it can only portray itself in itself, and describe itself. That, logically, is also the principle of my own method: it is, at bottom, a purely experiential process (1946b, § 421).

All he could do was "to compare individual psychic occurrences with obviously related collective phenomena" (1946b, § 436). "What I have practiced is simply a comparative phenomenology of the mind, nothing else. . . . *There is only one method: the comparative method*" (*Jung Speaking*, 1977, p. 220; my ital.). Consequently, the "comparative psychologist" cannot help but draw "even of the most obvious and superficial analogies, however fortuitous they may seem, because they serve as bridges for psychic associations" (1959, § 900).

At the heart of the matter is the fact that anything humans say about themselves is self-referential and lacks, as Jung stated numerous times, an outside "Archimedean point," from which objective conclusions could be drawn.[51] In other words, in psychology (as in philosophy) the observer and the observed coincide. This is a central point in Schopenhauer's philosophy:

> [E]ven in self-consciousness, the I is not absolutely simple, but consists of a knower (intellect) and a known (will); the former is not known and the latter is not knowing, although the two flow together into the consciousness of an I. But on this very account, this I is not *intimate* with itself through and through, does not shine through so to speak, but is opaque, and therefore remains a riddle to itself (1844 [1969], vol. 2, p. 196).

[51] A list of references in Jung & Schmid (2013), pp. 15–16.

Schopenhauer compared the human condition to a tree:

> [H]uman nature divides into *will* and *representation*, the former is the root, the latter the crown. *The I* is their point of indifference,[52] which unites the two and is part of both.... It is the point where the being-in-itself and its appearance meet: as an indivisible point it belongs in equal parts to the intellect and to the will, and this explains the miracle κατ' εξοχην,[53] namely, that that which wills, and that which cognizes, are one and the same (1966–1975 [1985], pp. 179–180).

"This 'I'—what a peculiar matter it is!" Jung wonders in these lectures. Well-versed in Kant and Schopenhauer, he saw the problem clearly: "The psyche [*Seele*] is the beginning and end of all cognition [*Erkennen*]. It is not only the object of its science, but the subject also. This gives psychology a unique place among all the other sciences: on the one hand there is a constant doubt as to the possibility of its being a science at all, while on the other hand psychology acquires the right to state a theoretical problem the solution of which will be one of the most difficult tasks for a future philosophy [*sic*!]" (1936 [1937], § 261).

As Schopenhauer had put it:

> [W]e are not merely the knowing subject, but ... we ourselves are the thing-in-itself. Consequently, a way from within stands open to us to that real inner nature of things to which we cannot penetrate from without. It is, so to speak, a subterranean passage, a secret alliance which, as if by treachery, places us all at once in the fortress that could not be taken by attack from without. Precisely as such, the thing-in-itself can come into consciousness only quite directly, namely by it itself being conscious of itself; to try to know it objectively is to desire something contradictory. Everything objective is representation, consequently appearance (1844 [1969], p. 195).

If we replace "thing-in-itself" with "unconscious," this could well have been written by Jung. Psychology as a conscious science is the effort of the soul/psyche to understand itself. It is "a mediatory science, and this alone is capable of uniting the idea and the thing without doing violence to either" (Jung, 1921, § 72). But the human condition itself, the "miracle κατ' εξοχην," the identity of the observer and the observed,

[52] That is, the "root stock, the point where they meet on the ground level" (ibid.).

[53] *kat' exochen*, Greek, par excellence.

and the very limits inherent in pure reason, seem to prevent final self-knowledge. Is there really a "subterranean passage" that could place us in the otherwise impenetrable fortress? Can psychology really be an "ordinary" empirical science? How can it avoid becoming Munchhausen psychology, pulling itself up by the bootstraps, so to speak, or endlessly going in circles? Or will it really be possible for psychology to become queen of all sciences, their basis, *fons et origo*, will indeed "Nietzsche be proved right in the end with his 'scientia ancilla psychologiae'?"[54] (Jung, 1930b, Introduction)

At times, Jung seemed to think that there is indeed an Archimedean point outside our self-referential system, from which to move the world of psychology. There would exist a "spiritual goal that points beyond the purely natural man and his worldly existence." Not only would such a spiritual goal be "an absolute necessity for the health of the soul," it would also represent "the Archimedean point from which alone it is possible to lift the world off its hinges" (1926 [1924], § 159). The unconscious is, by definition, not conscious, and as such not knowable. What may sound like a platitude is in fact the core problem of depth psychology. Does something like "the" unconscious exist at all? Or is this an inadmissible hypostazation? But if it exists, how can we ever expect to know it? Is this not a contradiction in itself?

Jung's "official" answer, and the stance he is taking in these lectures at the university, was always that he was simply dealing with psychical "facts," and that therefore he was an empiricist, period. It is a fact, for example, that some people experience ghosts, spirits, and even converse with them, or that they have visions. The difficulty, as he notes in these lectures, is just that the "only guarantee we have that such things do exist is the evidence of the I. People are confronted with them through the I, as if something existed behind the I whose source we are completely ignorant of." In other words: "I am the sole proof, for no one else has seen the event." Nevertheless, according to Jung, we have to take these reports, at least for the time being, at their face value. Some people do have such experiences, Jung repeats, and regardless of whether or not these experiences correspond to something in what others perceive as observable reality, they represent incontestable *psychical* facts and have to be taken as such by a psychology worthy its name.

---

[54] "Science [is] the handmaiden of psychology": a play on the frequent criticism of philosophy as the "handmaiden of theology" (*philosophia ancilla theologiae*). Nietzsche had demanded "that psychology again be recognized as queen of the sciences, and that the rest of the sciences exist to serve and prepare for it" (1886 [2002], p. 24).

Marilyn Nagy notes that "Jung struggled his whole life long to explain . . . that he did not intend to do philosophy and that 'the psyche is a phenomenal world in itself, which can be reduced neither to the brain nor to metaphysics.'" "Much confusion has arisen in the attempt . . . to understand Jung's description of himself as an empiricist and at the same time his insistence on the ultimate reality of psychic life" (1991, pp. 1, 20).

Sonu Shamdasani observes that Jung's attitude toward a purely scientific/empiricist approach, as opposed to a philosophical and metaphysical one, *changed* over time, notably after having chosen psychiatry as his profession:

> Between Jung's Zofingia lectures and his first publications, there are considerable discontinuities in language, conceptions, and epistemology, as the far-reaching speculations on metaphysical issues characteristic of the Zofingia lectures largely disappeared. Following his discovery of his vocation as a psychiatrist, *he appears to have undergone something like a conversion to a natural scientific perspective*. . . . [In 1900,] Jung stated that he would stand in for the standpoint of the natural sciences, where "one is accustomed to operate only with clear firmly defined concepts." He then launched on a critique of theology, religion, and the existence of God, which led one person to remark on the fact that Jung had previously held so many positive views on these subjects, which he had now abandoned (Shamdasani, 2003, p. 201; my ital.).

A crucial turning point in this "conversion" seems to have been his experimenting with Helene Preiswerk: "[T]his was the one great experience which wiped out all my earlier philosophy and made it possible for me to achieve a psychological point of view. I had discovered some objective facts about the human psyche" (*Memories*, p. 128).[55]

It can be questioned, however, if this change in orientation lasted. In *Memories*, Jung had talked about his "inner dichotomy" (p. 91), the "play and counterplay between personalities No. 1 and No. 2, which has run through my whole life" (p. 62). Although he was quick to assert that this had "nothing to do with a 'split' or dissociation in the ordinary medical sense," and that this pair of opposites "is played out in every individual" (ibid.), it seems safe to assume that this dichotomy

[55] For a detailed study of the impact of the séances with Helene Preiswerk, "that served as an impetus for my future life," as Jung recalled in 1925 (2012 [1925], p. 3), see Shamdasani (2015).

was particularly distinct in his case. Goethe's *Faust* was like a revelation: "*Faust* . . . pierced me through in a way that I could not but regard as personal. . . . Faust, the inept, purblind philosopher, encounters the dark side of his being, his sinister shadow, Mephistopheles, who in spite of his negative disposition represents the true spirit of life. . . . My own inner contradictions appeared here in dramatised form. . . . The dichotomy of Faust–Mephistopheles came together within myself into a single person . . . I was directly struck, and recognised that this was my fate" (p. 262). In Jung's theory, this dichotomy is reflected in his oscillating stance toward a philosophical, metaphysical approach versus a natural scientific perspective. Although psychiatry, and then psychology, seemed to offer a way out of this quandary, he remained caught up in it: He wanted to create the science of dreams, and ended up with the dream of a science.

Jung complained repeatedly about being constantly misunderstood, and that only a "chosen few" would be able to understand what he was aiming to convey: "There are only a few heaven-inspired minds who understand me" (*Jung Speaking*, p. 221). On the other hand, we also find frequent hints that there was actually more to the whole story, that he held back something, that he did not tell everything he knew, or seemed to know, so as not to be seen as "crazy," or even that the language that he used was deliberately obscure. It was his explicit "intention to write in such a way that fools get scared and only true scholars and seekers can enjoy its reading" (letter to Wilfred Lay, 20 April 1946; in Shamdasani, 2000). "The language that I use must be ambiguous or equivocal in order to do justice to the double aspect of psychical nature. I try deliberately and consciously to use ambiguous formulations, because they are superior to unambiguous ones, and correspond better to the nature of [our] existence" (*Memories,* only in German ed., p. 375).

"Everything profound loves masks," Nietzsche had written (1886 [2002], p. 38), "the most profound things go so far as to hate images and likenesses. Wouldn't just the opposite be a proper disguise for the shame of a god?" Was Jung one of those who are "hidden in this way," someone who "*wants* and encourages a mask of himself to wander around in his place" (ibid.)?

But does it really need obscure and ambiguous language to investigate and describe the complexity, the double-sided aspects of a topic? In these lectures, for example, Jung gave a lucid and absolutely clear description of what he called the "tremendous tension between the two poles" in the human condition.

Here is what Jung said about himself and his personality:

> I have intuitions about the subjective factor, the inner world. That is very difficult to understand because what I see are most uncommon things, and I don't like to talk about them because I am not a fool. I would spoil my own game by telling what I see, because people won't understand it. . . . So you see, if I were to speak of what I really perceive, practically no one would understand me. I have learned to keep things to myself, and you will hardly ever hear me talking of these things. That is a great disadvantage, but it is an enormous advantage in another way, not to speak of the experiences I have in that respect and also in my human relations. For instance, I come into the presence of somebody I don't know, and suddenly I have inner images, and these images give me more or less complete information about the psychology of the partner. It can also happen that I come into the presence of somebody I don't know at all, not from Adam, and I know an important piece out of the biography of that person, and am not aware of it, and I tell the story, and then the fat is in the fire. So I have in a way a very difficult life, although one of the most interesting lives, but it is often difficult to get into my confidence. [Interviewer:] *Yes, because you say you are afraid people will think you are sick.*] [Jung:] The things that are interesting to me, or are vital to me, are utterly strange to the ordinary individual (*Jung Speaking*, pp. 309–311).

A stunningly open self-diagnosis, one might say, if—well, if I had not taken the liberty of using a little ruse by replacing, in this quote, the term "intuitive introvert" by the first person singular. Still, Jung so often described himself as an intuitive introvert that I think we are justified in applying this description to himself, and that he might have even played a mischievous trick on his interviewer by providing some (again, veiled!) insight into his own personality.

These lectures, however, are a perfect counterexample to his intention to scare fools away, and can serve as a contrast (if we do not go so far as to see them as an example of how Jung "instinctively need[ed] speech in order to be silent and concealed" [Nietzsche, 1886 [2002], p. 38]). Here, Jung was very much concerned about being popular, that is, intelligible to all, speaking in layman's terms. "[Y]ou asked," a woman wrote him, "if your explanations would be popular enough," and she assured him that many found the lectures even "too popular"! Here, he was the university professor, having finally secured, after a long and difficult detour, a prestigious and coveted academic position. He tried to confront his audience with simple and observable "facts," but facts that were so strange and peculiar that he could thus prepare them for an acknowledgement and a discussion of a world much different from that which they readily acknowledged as "real."

To this end, Jung devoted much time in the second half to the discussion of two exemplary historical case studies, that of Friederike Hauffe (the "Seeress of Prevorst") and of Flournoy's medium Catherine-Elise Müller, aka "Hélène Smith." He took great pains to stress repeatedly that the psychic mechanisms that could be studied in these cases were by no means exceptional and to be found not only in such "border cases," but that they were universal. Since Hauffe and Müller exhibited certain traits and mechanisms to an extreme extent, however, those could be studied in isolation with the help of their cases. Nevertheless, they existed in every human being, including the members of his audience, as he did not tire of pointing out: "You are simply unaware that your own case exhibits all these basic facts, too, only they lie concealed in the dark background of your psyche. . . . The ideas that I have set forth in my lectures on the basis of this case have already been published, and I am not to blame if these are not more widely known!" Another reason for choosing these cases was that Jung had not been involved in them "in the least; otherwise, one would say again: 'Well, of course, he simply influenced the patient's mind!'" And again he stressed: "It is nothing other than an absolutely basic fact about the human soul; it is known all over the world, and, if we do not know it, then we are the morons!"

It is true that Jung interspersed his lectures with all kinds of anecdotes from his own practice, for example of the patient who "was so preoccupied with her psychological problems that she once sat down on a bench by the lake to dwell on them, although the thermometer showed minus six degrees Celsius. She sat there for two hours and was surprised that she had to pay for her folly with a severe cold, inflammation of the bladder, etc." Or of the neurotic "girl who had enjoyed the best education, and led an extremely sheltered life," but who when agitated would, "to the shock of her parents, . . . utter a flood of the most incredible expletives, on which even a wagoner could have prided himself." But he used these anecdotes only to illustrate points, often in an amusing way, which he had already made with the help of another case that was not his.

Similarly, we find quite a number of anecdotes about indigenous peoples he called "primitives," who would express clearly certain peculiarities of the human psyche that were often hidden in educated, "modern" man with his one-sided orientation. These remarks on "primitives" may strike us as rather condescending, or even racist, from a contemporary perspective, and although Jung also underlined the wisdom and perceptiveness of the "primitives"—as in the often quoted remark of "Mountain Lake" that the Americans are mad because they say they think with their heads—this may be felt as only highlighting the underlying conviction of their "primitiveness." We should bear in mind, however, that he was in good company, as it were, with such views at the time, and that our contemporary views, and also our terminology, have undergone radical changes since that era (and are probably also not set in stone).

With his strategy, Jung was able to kill two birds with one stone. First, he could present himself as an "objective" university professor, who gave a seemingly unbiased overview of the field, and a fair hearing to various different theories and systems of thought, thus fulfilling the exigencies of a university and of academic teaching. Second, at the same time he was able to set the background for his own views, and to firmly situate himself in a line of pre-eminent thinkers over the centuries. Hardly noticeable at first, but becoming ever more clear, this suggested that his theory was the culmination point, if only provisionally, of what so many great philosophers and psychologists—from Descartes through Leibniz, Locke, Hume, Kant, Hegel, Schopenhauer, Nietzsche etc. etc. up to Freud—had struggled with, while getting only glimpses of the big picture.

This account also provides a critical correction to Freudocentric accounts of the development of Jung's work, which were already in circulation at this time. The detailed taxonomy of consciousness that he presented toward the end of these lectures is certainly a highlight. Curiously enough, and although he claimed that this diagram was "the result of much deliberation and comparison" and "the fruit of encounters with people from all walks of life, from many countries and continents," it is not documented in any of his published works. The main question it addresses is: Where does the light of consciousness fall? In presenting it, Jung noted that the difficulties which he had encountered with his project for a psychological typology had led him to undertake this. Simply put, it is yet another attempt to explain the fact that people constantly do not understand, indeed misunderstand, one another. According to his typology, this is due, for instance, to one person being an introverted thinking type, and the other an extraverted feeling type. In this new classification, this would be because one person would live in III Right, and the other in IV Left, for example. And although changes of one's position in this diagram may occur individually over the course of one's lifetime, or historically with the advent of a new era, in general such viewpoints are "set in stone," as Jung put it, and "it is extremely rare that someone is willing to abandon the present position of his consciousness. Once consciousness has claimed a certain resting point, it can barely be shifted from its localization." Moreover, whereas *others* may see a potential in our position that we ourselves are unaware of, we are unable to understand their message: "An intuitive type, it is true, sees dozens of possibilities in other spheres, but he does not actually go there to experience them. For example, he sees a person living in Right IV as he appears to him from his vantage point in Left III. Consequently, the intuitive may see a great deal of which the man in Right IV is not aware, but what he says is unintelligible to the man himself because he does not know that Left III exists at all."

We may wonder why Jung never published this "result of much deliberation," all the more so in that this new "typology" would neither devalue nor exclude his former one, but would be, on the contrary, a perfect *complement* to it. As Jung also points out in these lectures, this new classification "refers exclusively to the shifts of consciousness, to its *localization*," whereas the typology of intro- and extraverts and of the four functions shows us "the quality of the personality that is the *bearer* of this consciousness."

Psychological Types (1921) had a long gestation time, of nearly a decade. In the wake of the original publication of *Transformations* (1911/1912), Jung tried to come to terms, not only with "the countless impressions and experiences of a psychiatrist," his "personal dealings with friend and foe alike," and the "critique of [his] own psychological peculiarity" (1921, p. xi), but also with the "dilemma" into which he was put by the differences between Freud's, Adler's, and his own theories. Now, again more than a decade later, he presented another classification, which can also be viewed as an attempt to answer precisely those questions.

We can only speculate about why Jung did not deem it necessary to present such a carefully thought-out classification in his published works. Did he think it was not important enough? Had he especially designed it for that specific audience, that is, for beginners, as something that might pave the way and prepare them for an understanding of his mature typology (to which he then indeed devoted two full semesters, the winter semester 1935/36 and the summer semester 1936)? Did he consider it a failure? Or did his interests at the time in general no longer focus on such classificatory attempts at all, and were already on to quite other things?

However that may be, here we encounter a Jung as we have not known him before. Not in his seminars, where admittance was strictly limited, participants had to seek personal permission from Jung to attend, and some basic familiarity with his concepts was taken for granted. Not in interviews given to popular newspapers or journals, and not in talks before various groups of laypersons, where he was much more limited in his time and possibilities. Here, however, before an audience of hundreds of people from all walks of life, in weekly meetings over several years, he could develop and lay out in detail, and in "popular" terms, the topics and concepts that were dear to him. I can only hope that the readers of this volume will enjoy this "unknown Jung" as much as I enjoyed preparing this text for publication.

Acknowledgments

THE PREPARATION FOR PUBLICATION OF THESE LECTURES, from thousands of pages of auditors' notes, has had a long gestation. Like a complex jigsaw puzzle assembled by numerous hands over many years, this work would not have been possible without the contributions of many individuals, to whom thanks are due. The Philemon Foundation, under its past presidents Steve Martin and Judith Harris, past copresident, Nancy Furlotti, and present president, Richard Skues, has been responsible for this project since 2004. Without the contributions of its donors, none of the editorial work would have been possible or come to fruition. From 2012, the project has been and continues to be supported by Judith Harris at UCL. From 2004 to 2011, the project was principally supported by Carolyn Fay, the C. G. Jung Educational Center of Houston, the MSST Foundation, and the Furlotti Family Foundation. The project was also supported by research grants from the International Association for Analytical Psychology in 2006, 2007, 2008, and 2009.

This publication project was commenced by the former Society of Heirs of C. G. Jung (now the Foundation of the Works of C. G. Jung), between 1993 and 1998. Since its inception, Ulrich Hoerni has been involved in nearly every phase of the project, actively supported between 1993 and 1998 by Peter Jung. The executive committee of the Society of Heirs of C. G. Jung released the scripts for publication. At ETH Zurich, the former head of the archives, Beat Glaus, made scripts available and supervised transcriptions. Ida Baumgartner and Silvia Bandel transcribed shorthand notes of the lectures; C. A. Meier provided general information about the lectures; Marie-Louise von Franz provided information about the editing of Barbara Hannah's scripts; Helga Egner and Sonu Shamdasani gave editorial advice; at the Jung Family Archives, Franz Jung and Andreas Jung made scripts and related materials available; at the Archives of the Psychological Club, the former chairman, Alfred Ribi, and the librarian, Gudrun Seel, made lecture notes available; Sonu Shamdasani found notes taken by Lucy Stutz-Meyer. Rolf Auf der Maur and Leo La Rosa provided legal advice and managed contracts.

In 2004, the Philemon Foundation took on the project, in collaboration with the Society of Heirs of C. G. Jung, and since 2007, with its successor organization, the Foundation of the Works of C. G. Jung, and the ETH Zurich Archives. At the Foundation of the Works of C. G. Jung, Ulrich Hoerni, former president and executive director, Daniel Niehus, president, and Thomas Fischer, executive director, oversaw the project, and Ulrich Hoerni, Thomas Fischer, and Bettina Kaufmann, editorial assistant, reviewed the manuscript. Since 2007, Peter Fritz of the Paul & Peter Fritz Agency has been responsible for managing contracts. At the ETH Zurich Archives, Rudolf Mumenthaler, Michael Gasser, former directors, Christian Huber, director, and Yvonne Voegeli made scripts and related documents available. Nomi Kluger-Nash provided Rivkah Schärf's shorthand notes of some of the lectures, which were then transcribed by Silvia Bandel. Steve Martin provided Bertha Bleuler's shorthand notes of some of the lectures.

The editorial work has been overseen by Sonu Shamdasani, general editor of the Philemon Foundation. Between 2004 and 2011, the preparatory phase of the compilation of the scripts and editorial work was undertaken by Angela Graf-Nold at the former Institute for the History of Medicine at the University of Zurich. From 2012 the compilation and editorial work has been undertaken by Ernst Falzeder and Martin Liebscher at the Health Humanities Centre and German department at UCL.

The editor of this volume, Ernst Falzeder, offers thanks to the board members of the Philemon Foundation, with particular gratitude to Judith Harris; Angela Graf-Nold, for providing the preparatory ground he could build on; Mark Kyburz and John Peck, who established a first translation on the basis of Graf-Nold's preliminary draft; the team at Princeton University Press, who were always supportive and always there in the sometimes difficult and delayed making of this book, in particular, Fred Appel, executive editor, Jay Boggis, copyeditor, Karen Carter, project manager, who dedicated much time to this project, and Virginia Ling, who created the index; Gertrude Enderle-Burcel, Österreichische Gesellschaft für historische Quellenstudien; Erika Gonsa; Thomas Fischer; Ulrich Hoerni; Martin Liebscher; Christine Maillard, University of Strasbourg; Sonu Shamdasani; Tony Woolfson; Gemmo Kosumi; Gerhard Laber; Marina Leitner; the members of the Phanês group (https://phanes.live/): Anna Dadaian, Alessio De Fiori, Gaia Domenici, Matei Iagher, Armelle Line-Peltier, Tommaso A. Priviero, Quentin Schaller, Florent Serina, Josh Torabi, and Dangwei Zhou; and the community of the translators' forum at http://dict.leo.org/forum/.305938ONA_.

Abbreviations

Principal Bibliographical References

CW = *The Collected Works of C. G. Jung*. Ed. Sir Herbert Read, Michael Fordham, Gerhard Adler. Trans. R. F. C. Hull. Princeton: Bollingen Series, Princeton University Press, 1953–1983. 21 vols.

Hannah = *Barbara Hannah (1938 [1959]). Modern Psychology,* Vols. 1 and 2: *Notes on Lectures Given at the Eidgenössische Technische Hochschule, Zürich, by Prof. Dr. C. G. Jung. October 1933—July 1935*. Second edition 1959. Zurich: mimeographed typescript.

Jung Speaking = *C. G. Jung Speaking: Interviews and Encounters*. Ed. William McGuire and R. F. C. Hull. Bollingen Series XCVII. Princeton: Princeton University Press, 1977.

Memories = C. G. Jung *Memories, Dreams, Reflections*. Recorded and edited by Aniela Jaffé, 1962. Trans. Richard and Clara Winston. London: Fontana Press, 1995.

Planet Mars = Théodore Flournoy (1900 [1899]). *Des Indes à la planète Mars. Étude sur un cas de somnambulisme avec glossolalie*. Paris, Geneva: F. Alcan, Ch. Eggimann. *From India to the Planet Mars, a Study of Somnambulism with Glossolalia. Die Seherin von Genf*. Foreword by Max Dessoir. Authorized translation. Leipzig: Felix Meiner Verlag, 1914. *From India to the Planet Mars: A Case of Multiple Personality with Imaginary Languages*. Foreword by C. G. Jung and Commentary by Mireille Cifali. Trans. Daniel B. Vermilye. Ed. and introduced by Sonu Shamdasani. Princeton: Princeton University Press, 1994.

Protocols = Protocols of Aniela Jaffé's interviews with Jung for *Memories, Dreams, Reflections*; Library of Congress, Washington DC (in German).

Seeress = Justinus Andreas Christian Kerner (1829). *Die Seherin von Prevorst. Eröffnungen über das innere Leben und über das Hineinragen einer Geisterwelt in die unsere.* Two vols. Stuttgart, Tubingen: J. G. Cotta'sche Buchhandlung, 1829. Fourth, expanded and revised edition. Stuttgart, Tubingen: J. G. Cotta'scher Verlag, 1846. Reprint: Kiel: J. F. Steinkopf Verlag, 2012. *The Seeress of Prevorst, Being Revelations Concerning the Inner-Life of Man, and the Inter-Diffusion of a World of Spirits in the One We Inhabit.* Trans. Catherine Crowe. London: J. C. Moore, 1845. Digital reprint: Cambridge: Cambridge University Press, 2011.

Transformations = Jung, C. G. (1911/12). "Wandlungen und Symbole der Libido." *Jahrbuch für psychoanalytische und psychopathologische Forschungen* 3(1) (1911): 120–227; 1912, 4(1): 162–464. In book form: Leipzig: Deuticke, 1912. Reprint: Munich: Deutscher Taschenbuch Verlag, 1991. In revised form (1950) and under new title, *Symbole der Wandlung*, in *GW* 5; *Symbols of Transformation*, in *CW* 5.

Types = Jung, C. G. *Psychologische Typen* (1921). *GW* 6. *Psychological Types*. *CW* 6.

Chronology 1933–1941

COMPILED BY ERNST FALZEDER, MARTIN LIEBSCHER, AND SONU SHAMDASANI

Date	*Events in Jung's Career*	*World Events*
1933		
January	Jung continues his English seminar on Christiana Morgan's visions, on Wednesday mornings.	
30 January		Hitler is appointed Reich Chancellor in Germany by the president, Paul von Hindenburg.
February	Jung lectures in Germany (Cologne and Essen) on "The Meaning of Psychology for Modern Man" (*CW* 10).	
27 February		Reichstag fire in Berlin. The fire, possibly a false flag operation, was used as evidence by the Nazis that the Communists were plotting against the German government, and the event is seen as pivotal in the establishment of Nazi Germany. Many arrests of leftists. On 28 February, the most important basic rights of the Weimar republic were suspended.
4 March		"Self-dissolution" of the Austrian parliament, and authoritarian régime under Chancellor Engelbert Dollfuß.
5 March		In the German federal elections, the National Socialists become the strongest party with 43.9 % of the votes.

Date	*Events in Jung's Career*	*World Events*
13 March to 6 April	Jung accepts the invitation of Hans Eduard Fierz to accompany him on a cruise on the Mediterranean, including a visit to Palestine.	
18/19 March	Athens. Visits the Parthenon and the theatre of Dionysus.	
23 March		The German parliament passes the *Ermächtigungsgesetz* (Enabling Act), according to which the government is empowered to enact laws without the consent of the parliament or the president of the Reich—a self-disempowerment of the parliament.
25–27 March	Jung and Fierz visit Jerusalem, Bethlehem, and the Dead Sea.	
28–31 March	Egypt, with visits to Gizeh and Luxor.	
March to June		Franklin D. Roosevelt starts the New Deal.
1 April		Nationwide boycott of Jewish shops in Germany.
5 April	Via Corfu and Ragusa the *General von Steuben* lands in Venice, from where Jung and Fierz take the train to Zurich.	
6 April	Ernst Kretschmer resigns from the presidency of the International General Medical Society for Psychotherapy (IGMSP) in protest against "political influences." Jung, as vice-president, accepts the acting presidency and editorship of the society's journal, the *Zentralblatt für Psychotherapie*.	
7 April		The German parliament passes a law that excludes Jews and dissidents from civil service.

Date	*Events in Jung's Career*	*World Events*
22 April		"Non-Aryan" teachers are excluded from their professional organizations, "non-Aryan" and "Marxist" physicians lose their accreditation with the national health insurance.
26 April		Formation of the Gestapo.
1–10 May		Ban on trade unions in Germany.
10 May		Public burning of books in Berlin and other cities, including those of Freud.
14 May	The Berliner *Börsen-Zeitung* publishes "Against psychoanalysis," describing Jung as the reformer of psychotherapy.	
22 May		Sándor Ferenczi dies in Budapest.
27 May/ 1 June		The German government imposes the so-called Thousand Mark Ban, an economic sanction against Austria. German citizens had to pay a fee of 1000 Reichsmark (or the equivalent of about $5,000 in 2015) to enter Austria.
21 June	Jung accepts the presidency of the IGMSP.	
26 June	Interview with Jung on Radio Berlin, conducted by Adolf Weizsäcker.	
26 June– 1 July	Jung gives the "Berlin Seminar," opened by a lecture by Heinrich Zimmer on 25 June.	
14 July		"Law for the prevention of hereditarily diseased offspring" in Germany, which allows the compulsory sterilization of any citizen with alleged hereditary diseases.
14 July		In Germany, all parties with the exception of the NSDAP are banned or dissolve themselves.

Date	*Events in Jung's Career*	*World Events*
August	Jung's first attendance at the Eranos meeting in Ascona, giving a talk on "On the Empirical Knowledge of the Individuation Process" (retitled, *CW* 9/1).	
15 September	Foundation of a new German chapter of the IGMSP, whose statutes demand unconditional loyalty to Hitler. Matthias H. Göring, a cousin of Hermann Göring, is named its president.	
22 September		Law on the "Reich chamber of culture" in Germany, enforced conformity [*Gleichschaltung*] of culture in general, tantamount to an occupational ban on Jews and artists who produce "degenerate" art.
7/8 October	Meeting of the Swiss Academy of Medical Science at Prangins. Jung presents a contribution on hallucination (*CW* 618).	
20 October	Jung's first lecture on "Modern Psychology" at ETH.	
5 December		Repeal of Prohibition in the United States with the passage of the Twenty-first Amendment.
10 December		Nobel Prize in Physics to Erwin Schrödinger and Paul A. M. Dirac "for the discovery of new productive forms of atomic theory."
December	Jung publishes an editorial in the *Zentralblatt* of the IGMSP, in which he contrasts "Germanic" with "Jewish" psychology (*CW* 10). The same issue contains a manifesto of Nazi principles by Matthias Göring that, be it by oversight or on purpose, also appears in the international, not only German, edition, against Jung's wishes. Jung threatens to resign from the presidency, but ultimately stays on.	

Date	Events in Jung's Career	World Events
Other Publications in 1933:		
"Crime and Soul," *CW* 18		
"On Psychology," revised version in *CW* 8		
"Brother Klaus," *CW* 11		
Foreword to Esther Harding, *The Way of All Women*, *CW* 18		
Review of Gustav Richard Heyer *Der Organismus der Seele*, *CW* 18		
1934		
20 January		German "Work Order Act" and introduction of the "Führer principle" in economy.
12–16 February		Civil war in Austria, resulting in a ban of all social-democratic parties and organizations, mass arrests, and summary executions.
23 February	Jung's last lecture at ETH in the winter semester of 1933/34.	
27 February	Gustav Bally publishes a letter to the editor of the *Neue Zürcher Zeitung* ("Psychotherapy of German Origin?"), in which he strongly criticizes Jung for his alleged Nazi leanings and anti-Semitic views.	
Spring	Beginning of Jung's serious and detailed study of alchemy, assisted by Marie-Louise von Franz.	
13–14 March	Jung publishes a rejoinder to Bally in the *NZZ* ("Contemporary Events", *CW* 10).	
16 March	Publication of B. Cohen, "Is C. G. Jung 'Conformed'?" in *Israelitisches Wochenblatt für die Schweiz*.	
21 March	Jung's last seminar on Christiana Morgan's visions. The participants opt for continuing the English Wednesday morning seminars with one on Nietzsche's *Zarathustra*.	

Date	Events in Jung's Career	World Events
March/April	C. G. Jung, *The Reality of the Soul: Applications and Advances of Modern Psychology*; with contributions from Hugo Rosenthal, Emma Jung, and W. M. Kranefeldt.	
April	Jung publishes "Soul and Death" (*CW* 8).	
April	Interview with Jung, "Does the World Stand on the Verge of Spiritual Rebirth?" (*Hearst's International-Cosmopolitan*, New York).	
ca. April	Jung publishes "On the Present Position of Psychotherapy" in the *Zentralblatt* (*CW* 10).	
20 April	Jung's first ETH lecture in the summer semester.	
2 May	Jung starts the English seminar on Nietzsche's *Zarathustra* (until 15 February 1939).	
5 May	Jung's inaugural lecture at ETH, "A General Review of Complex Theory" (*CW* 8).	
10–13 May	Jung presides at the Seventh Congress for Psychotherapy in Bad Nauheim, Germany, and repeats his talk on the complex theory. Foundation of an international umbrella society for the IGMSP, organized in national groups that are free to make their own regulations. On Jung's proposition statutes are passed that (1) provide that no single national society can muster more than 40 % of the votes, and (2) allow that individuals (that is, Jews, who are banned from the German society) can join the International Society as "individual members." Jung is confirmed as president and as editor of the *Zentralblatt*.	

Date	*Events in Jung's Career*	*World Events*
29 May	James Kirsch, "The Jewish Question in Psychotherapy: A Few Remarks on an Essay by C. G. Jung," in the *Jüdische Rundschau*.	
31 May		The "Barmen Declaration," mainly instigated by Karl Barth, openly repudiates the Nazi ideology. It becomes one of the founding documents of the Confessing Church, the spiritual resistance against National Socialism.
15 June	Erich Neumann, letter to the *Jüdische Rundschau* regarding Kirsch's "The Jewish Question in Psychotherapy."	
30 June/ 1 July		The so-called Röhm putsch. SA leader Ernst Röhm, other high-ranking SA members, and alleged political opponents are executed on Hitler's direct orders, among them Röhm's personal physician Karl-Günther Heimsoth, a longtime member of the IGMSP and a personal acquaintance of Jung.
13 July	Jung's last ETH lecture in the summer semester.	
25 July		Failed putsch attempt by the Nazis in Austria, in which the Austrian chancellor Engelbert Dollfuß is murdered.
29 July		New government in Austria under chancellor Kurt Schuschnigg who tries to control the Nazi movement by his own authoritarian, right-wing regime.
2 August		Death of Reich president Paul von Hindenburg. Hitler assumes chancellorship and presidency in personal union, as well as supreme command of the Wehrmacht.
3 August	Gerhard Adler, "Is Jung an Antisemite?", in the *Jüdische Rundschau*.	

Date	Events in Jung's Career	World Events
August	Eranos meeting in Ascona. Jung talks on "The Archetypes of the Collective Unconscious" (*CW* 9/1).	
1–7 October	Jung gives a seminar at the Société de Psychologie in Basle.	
26 October	First ETH lecture of the winter semester 1934/35.	

Other publications in 1934:

With M. H. Göring, "Geheimrat Sommer on his 70th Birthday," *Zentralblatt* VII

Circular letter, *Zentralblatt*, *CW* 10

Addendum to "Zeitgenössisches," *CW* 10

Foreword to Carl Ludwig Schleich, *Die Wunder der Seele*, *CW* 18

Foreword to Gerhard Adler, *Entdeckung der Seele*, *CW* 18

Review of Hermann Keyserling, *La Révolution Mondiale*, *CW* 10

1935

Date	Events in Jung's Career	World Events
	Jung becomes titular professor at ETH.	
	Jung completes his tower at Bollingen, by adding a courtyard and a loggia.	
19 January	Jung accepts an invitation to lecture in Holland.	
22 January	Foundation of the Swiss chapter of the IGMSP.	
24 February		Swiss extend the period of military training.
1 March		Saarland reunion with Germany, marking the beginning of German expansion under the National Socialists.
8 March	Final ETH lecture of the winter semester 1934/35.	

Date	*Events in Jung's Career*	*World Events*
16 March		The German government officially denounces its future adherence to the disarmament clauses of the Versailles Treaty.
26 March		Switzerland bans slanderous criticisms of state institutions in the press.
27–30 March	Eighth Congress of the IGMSP in Bad Nauheim (*CW* 10).	
2 May		Franco-Russian Alliance.
3 May	First ETH lecture of the Summer semester 1935.	
May	Jung attends and lectures at an IGMSP symposium on Psychotherapy in Switzerland.	
5 June		The Swiss government introduces an extensive armament expansion program.
11 June		The disarmament conference in Geneva ends in failure.
28 June	Publication of Jung's contribution at the May IGMSP symposium, "What Is Psychotherapy?", in the *Schweizerische Ärztezeitung für Standesfragen* (*CW* 16).	
12 July	Jung's last ETH lecture in the summer semester.	
August	Eranos lecture on "Dream Symbols of the Individuation Process" (*CW* 9/1).	
15 September		Passing of the so-called Nuremberg Laws in Germany. These laws deprive Jews (defined as all those one-quarter Jewish or more) and other non-"Aryans" of German citizenship, and prohibit sexual relations and marriages between Germans and Jews.

Date	Events in Jung's Career	World Events
30 September–4 October	Jung gives five lectures at the Institute of Medical Psychology in London, to an audience of around one hundred (*CW* 18).	
October		Conclusion of the "Long March" in China.
2 October	Publication of Jung's "The Psychology of Dying" (a shortened version of "Soul and Death") in the *Münchner Neueste Nachrichten* (*CW* 8).	
2–3 October		Italian invasion of Ethiopia.
25 October	First ETH lecture of the Winter semester 1935/36.	
6 October	Interview with Jung, "Man's immortal mind," *The Observer*.	
8 November		Switzerland tightens banking secrecy laws (leading to the numbered bank accounts).
December		Nobel Peace Prize for leftist German journalist and editor Carl von Ossietzky. Hitler forbids Germans to accept Nobel Prizes.
15 October	The Dutch national group of the IGMSP retracts their invitation to host its next international congress, because of the events in Nazi Germany. In his answer, Jung states that this "compromises the ultimate purpose of our international association," and declares that he will resign as its president, which he does not carry through, however.	

Date	Events in Jung's Career	World Events
Other publications in 1935:		
The Relations between the I and the Unconscious, 7th edition, CW 7		
Introduction and psychological commentary on the *The Tibetan Book of the Dead*, CW 11		
"Votum C.G. Jung", CW 10		
"Editorial" (*Zentralblatt* VIII), CW 10		
"Editorial Note" (*Zentralblatt* VIII), CW 10		
"Fundamentals of Practical Psychotherapy", CW 16		
Foreword to Olga von Koenig-Fachsenfeld, *Wandlungen des Traumproblems von der Romantik bis zur Gegenwart*, CW 18		
Foreword to Rose Mehlich, *J. H. Fichtes Seelenlehre und ihre Beziehung zur Gegenwart*, CW 18		
1936		
February	"Yoga and the West" (CW 11).	
February	"Psychological Typology" (CW 6).	
27 February		Death of Iwan Pawlow.
Spring	Formation of the Analytical Psychology Club in New York City.	
March	Jung publishes "Wotan" in the *Neue Schweizer Rundschau* (CW 10).	
6 March	Final ETH lecture of the winter semester 1935/36.	
7 March		German military forces enter the Rhineland, violating the terms of the Treaty of Versailles and the Locarno Treaties. This remilitarization changes the balance of power in Europe from France towards Germany.
28 March		The property of the *Internationaler Psychoanalytischer Verlag*, and all its stock of books and journals, are confiscated.

Date	*Events in Jung's Career*	*World Events*
May		Foundation of the *Deutsches Institut für psychologische Forschung und Psychotherapie* in Berlin, headed by M. H. Göring ("Göring Institute"), with working groups of Jungian, Adlerian, and Freudian orientation. Psychoanalysis was tolerated, but on the condition that its terminology be altered.
May	"Concerning the Archetypes, With Special Consideration of the Anima Concept," in the *Zentralblatt* (CW 9/1).	
1 May	First ETH lecture of the summer semester 1936.	
July		Beginning of Spanish civil war.
10 July	Final ETH lecture of the summer semester 1936.	
19 July	Jung and Göring attend a meeting of psychotherapists in Basel, with representatives of different depth-psychological schools (among others, Ernest Jones for the International Psycho-Analytical Association).	
August	Eranos meeting; Jung speaks on "Representations of Redemption in Alchemy" (CW 12).	
1–16 August		Summer Olympics in Berlin. Germans who are Jewish or Roma are virtually barred from participating.
21–30 August	Jung travels on board the *Georgic* from Le Havre to New York City. Upon arrival in New York, he releases a "Press Communiqué on Visiting the United States," setting forth his political—or, as he insisted, his nonpolitical—position.	

Date	*Events in Jung's Career*	*World Events*
September	Jung lectures at the Harvard Tercentenary Conference on Arts and Sciences, on "Psychological Factors Determining Human Behavior" (*CW* 8), and receives an honorary degree. His invitation had given rise to controversy.	
12–15 September	Jung is guest of the Anglican bishop James De Wolf Perry in Providence, Rhode Island, addresses the organization "The American Way," and then leaves for Milton, Mass., where he is guest of G. Stanley Cobb.	
ca. 19 September	Jung starts a seminar on Bailey Island, based on Wolfgang Pauli's dreams.	
2 October	Jung gives a public lecture at the Plaza Hotel in NYC. The talk is privately published by the New York Analytical Psychology Club under the title, "The Concept of the Collective Unconscious." (*CW* 9/1).	
3 October	Jung leaves New York City.	
4 October	Interview with Jung, "Roosevelt 'Great,' Is Jung's Analysis," *New York Times* (later published under the title, "The 2,000,000-year-old-man").	
14 October	Jung lectures at the Institute of Medical Psychology, London, on "Psychology and National Problems" (*CW* 18).	
15 October	Interview with Jung, "Why the World Is in a Mess. Dr. Jung Tells Us how Nature Is Changing Modern Woman," *Daily Sketch*.	
18 October	Interview with Jung, "The Psychology Of Dictatorship," *The Observer*.	

Date	Events in Jung's Career	World Events
19 October	Jung lectures before the Abernethian Society, St. Bartholomew's Hospital, London, on the concept of the collective unconscious (*CW* 9/1).	
25 October		Secret peace treaty between Germany and Italy.
27 October	Jung begins his seminars at ETH on children's dreams and old books on dream interpretation.	
3 November		Franklin D. Roosevelt is re-elected for his second term.
25 November		Anti-Comintern Pact between Germany and the Empire of Japan, directed against the Third (Communist) International.
10 December		Abdication of Edward VIII in England.

Other publications in 1936:

Review of Gustav Richard Heyer, *Praktische Seelenheilkunde*, *CW* 18

1937		
3–5 January	Jung participates in the workshop of the Köngener Kreis (1–6 January) in Königsfeld (Black Forest, Germany), on "Grundfragen der Seelenkunde und Seelenführung" [Fundamental Questions of the Study and Guidance of the Soul].	
30 January		Hitler formally withdraws Germany from the Versailles Treaty. This includes Germany no longer making reparation payments. He demands a return of Germany's colonies.
23 April 1937	After a break in the winter semester Jung's ETH lectures commence.	

Date	*Events in Jung's Career*	*World Events*
26 April		Germany and Italy are allied with Franco and the fascists in Spain. German and Italian airplanes bomb the city of Guernica, killing more than 1,600.
23 May		Death of John D. Rockefeller.
28 May		Death of Alfred Adler in Aberdeen, Scotland.
9 July	Final ETH lecture of the summer semester 1937.	
19 July		The NS exhibition on "Degenerate art" opens at the Institute of Archaeology, Munich.
August	Eranos Lecture on "The Visions of Zosimos" (*CW* 13).	
2–4 October	Ninth International Medical Congress for Psychotherapy in Copenhagen, under the presidency of Jung (*CW* 10).	
October	Jung is invited to Yale University to deliver the fifteenth series of "Lectures on Religion in the Light of Science and Philosophy" under the auspices of the Dwight Harrington Terry Foundation (published as "Psychology and Religion," *CW* 11).	
	Dream Seminar (continuation from the Bailey Island seminars), Analytical Psychology Club, New York.	
December	Jung is invited by the British Government to take part in the celebrations of the 25th anniversary of the founding of the Indian Science Congress Association at the University of Calcutta. He is accompanied by Harold Fowler McCormick Jr. (1898–1973) and travels through India for three months.	

Date	Events in Jung's Career	World Events
13 December		Nanjing falls to the Japanese. In the six weeks to follow, the Japanese troops commit war crimes against the civilian population known as the Nanjing Massacre.
17 December	Arrival in Bombay by P & O Cathay.	
19 December	Jung reaches Hyderabad, where he is bestowed an Honorary Doctor Degree by the University Osmania in Hyderabad; night train to Aurangabad.	
20 December	Aurangabad: visits the Kailash Temple at Ellora, and Daulatabad.	
21 December	Visits the caves at Ajanta.	
22 December	Sanchi, Bhopal, visits the Great Stupa.	
23 December	Taj Mahal, Agra.	
27 December	Benares; Jung visits Sarnath.	
28 December	Jung is awarded the D. Litt. (Doctor of Letters) Honoris Causa by the Benares Hindu University; presentation at the Philosophy Department: "Fundamental Conceptions Of Analytical Psychology"; guest of Swiss interpreter of Indian Art Alice Boner; visits the Vishvanatha Śiva Temple.	
29 December	Calcutta.	
31 December	Jung travels to Darjeeling.	

Other publications in 1937:

"On the Psychological Diagnosis of Facts: The Fact Experiment in the Näf Court Case," CW 2

Date	Events in Jung's Career	World Events
1938		
1 January	Three-hour conversation with Rimpotche Lingdam Gomchen at the Bhutia Busty monastery.	
3 January	Opening of the 25th anniversary of the founding of the Indian Science Congress Association at the University of Calcutta.	
	Jung is treated in the hospital in Calcutta.	
7 January	Jung is awarded (in absentia) the degree of Doctor of Law (Honoris Causa) by the University of Calcutta.	
10 January	Lecture at the College of Science, University of Calcutta: "Archetypes of the collective unconscious."	
11 January	Lecture at the Ashutosh College, University of Calcutta: "The Conceptions of Analytical Psychology."	
13 January	Visits the Temple of Konark ("Black Pagoda").	
21 January	Visits the Chennakesava Temple (also called the Kesava temple) and the temple of Somanathapur (Mysore).	
26 January	Jung in Trivandrum; lecture at the University of Travancore: "The Collective Unconscious."	
27 January	University of Travancore: "Historical Developments of the Idea of the Unconscious."	
28 January	Ferry to Ceylon.	
29 January	Colombo.	

Date	Events in Jung's Career	World Events
30 January	Train to Kandy.	
1 February	Return to Colombo.	
2 February	Embarks on the S.S. *Korfu* to return to Europe.	
12 March		Annexation of Austria by Nazi-Germany.
27 April		Edmund Husserl, the founding philosopher of phenomenology, dies in Freiburg, Germany.
May		The League of Nations acknowledges the neutral status of Switzerland.
29 April	After his return from India, Jung's ETH lecture series recommences.	
4 June		Sigmund Freud leaves Vienna; after a stop in Paris he arrives in London two days later.
8 July	Final ETH lecture of the summer semester 1938.	
29 July–2 August	Tenth International Medical Congress for Psychotherapy in Balliol College, Oxford, under the presidency of Jung; honorary doctorate from the University of Oxford. "Presidential Address" (*CW* 10).	
August	Eranos Lecture on "Psychological Aspects of the Mother Archetype" (*CW* 9/1).	
29 September		Munich Pact permits Nazi Germany the immediate occupation of the Sudentenland.
		Agreement between Switzerland and Germany concerning the stamping of German Jewish passports with "J."
28 October	First ETH lecture of the winter semester 1938/39.	

Date	Events in Jung's Career	World Events
October	Jung's ETH seminar series on the psychological interpretation of children's dreams commences in the winter term of 1938/39.	
9 November		A Swiss theology student, Maurice Bauvaud, fails to assassinate Hitler at a Nazi parade in Munich, and is guillotined.
9/10 November		Pogrom against Jews in Nazi Germany ("Crystal Night").
23 November	Jung gives his witness statement at the retrial of the murder case of Hans Näf.	

Other publications in 1938:

With Richard Wilhelm, *The Secret of the Golden Flower*, 2nd edition, *CW* 13

"On the *Rosarium Philosophorum*", *CW* 18

Foreword to Gertrud Gilli, *Der dunkle Bruder*, *CW* 18

1939

Date	Events in Jung's Career	World Events
January 1939	"Diagnosing the dictators," interview with H. L. Knickerbocker, *Hearst's International-Cosmopolitan.*	
15 February	The last of Jung's seminars on Nietzsche's *Zarathustra*, and hence of Jung's regular English-language seminars.	
3 March	Final ETH lecture of the winter semester 1938/39.	
28 March		Madrid surrenders to the Nationalists; Franco declares victory on 1 April.
April	Visits the west country in England in connection with Emma Jung's Grail research.	
4 April	Lecture at the Royal Society of Medicine in London, "On the Psychogenesis of Schizophrenia" (*CW* 3).	

Date	*Events in Jung's Career*	*World Events*
5 April	Lecture at the Guild of Pastoral Psychology, London, on "The Symbolic Life."	
28 April	First ETH lecture of the summer semester 1939.	
May	Surendranath Dasgupta lectures on Patanjali's *Yoga Sutras* in the Psychology Club Zurich.	
	Interview with Howard Philp, "Jung Diagnoses the Dictators," *Psychologist*.	
July	At a meeting of delegates of the International General Medical Society for Psychotherapy Jung offers his resignation.	
7 July	Final lecture of the summer semester 1939.	
August	Eranos Lecture on "Concerning Rebirth" (*CW* 9/1).	
1 September		Nazi-German troops invade Poland; Britain and France declare war on Germany two days later; begin of World War II.
		Switzerland proclaims neutrality.
23 September		Sigmund Freud dies in London at the age of 83.
	Moves his family for safety to Saanen in the Bernese Oberland.	
1 October	Jung's obituary of Freud is published in the *Sonntagsblatt der Basler Nachrichten* (*CW* 15).	
3 October	First ETH lecture of the winter semester 1939/40.	

Date	Events in Jung's Career	World Events
October	Jung's ETH seminar series on the psychological interpretation of children's dreams commences in the winter term 1939/40.	

Other publications in 1939:

"Consciousness, Unconscious and Individuation," *CW* 9/1

"The Dreamlike World of India" and "What India Can Teach Us," *CW* 10

Foreword to Daisetz Teitaro Suzuki's *Introduction to Zen-Buddhism*, *CW* 10

1940		
8 March	Final ETH lecture of the winter semester 1939/40.	
9 April		German troops invade Norway and Denmark.
10 May		German invasion of Belgium, the Netherlands, and Luxembourg.
12 May		France is invaded by Germany.
14 June		German troops occupy Paris.
20 June	In a letter to Matthias Göring, Jung offers his resignation of the presidency of the International General Medical Society for Psychotherapy.	
12 July	Jung sends his final letter of resignation to M. Göring.	
19 July		Hermann Göring is appointed Reichsmarschall.
August	Eranos lecture on "A psychological approach to the dogma of the trinity" (*CW* 11).	
7 September (–21 May 1941)		German aerial raids against London ("the Blitz").

Date	Events in Jung's Career	World Events
29 October	Jung's ETH seminar series on children's dreams commences in the winter semester 1940/41.	
8 November	First ETH lecture of the winter semester 1940/41.	

Other publications in 1940:

Foreword to Jolande Jacobi, *Die Psychologie von C. G. Jung*, CW 18

1941

Date	Events in Jung's Career	World Events
13 January		Death of James Joyce in Zurich.
28 February	Final lecture of the winter semester 1940/41.	
2 May	First ETH lecture of the summer semester 1941.	
11 July	Jung's final ETH lecture.	
August	Eranos lecture on "Transformation Symbolism in the Mass" (*CW* 11).	
7 September	Presents a lecture on "Paracelsus as a Doctor" to the Swiss Society for the History of Medicine in Basel (*CW* 15).	
5 October	Presents a lecture on "Paracelsus as a Spiritual Phenomenon" in Einsiedeln, on the 400th anniversary of the death of Paracelsus (*CW* 13).	

Other publications in 1941:

Essays on a Science of Mythology. The Myth of the Divine Child and the Mysteries of Eleusis, together with Karl Kerényi, CW 9/1

"Return to the Simple Life," CW 18

History of Modern Psychology

Lecture 1

20 October 1933

Twenty years ago, I resigned from my lectureship at the university. At the time, I had been lecturing for eight years, of course with mixed success. Eventually, I realized that one must understand something about psychology in the first place before being able to lecture about it.[56] I then withdrew, and travelled the world, since our cultural sphere simply fails to supply us with an Archimedean point.[57]

Now, after twenty years of professional experience, I am returning to the lecture hall, and will attempt to convey to you a sense of the field known as "psychology." By no means is this a simple undertaking, as I am sure you will agree. It is very difficult to present such a comprehensive field in a generally intelligible and somewhat concentrated manner, particularly since it occupies such an incredibly vast area. The human soul is enormously complicated, and about as many psychologies could be written as there are minds. Some psychologies address highly specific questions, such as those pertaining to biology or to the individual.

Each year, Clark University in Worcester, Massachusetts, publishes a weighty tome five centimeters thick and entitled *Psychologies of 1933*, etc.[58] I must therefore chart a path through this incredible chaos of opinions. I have

---

[56] As a *Privatdocent* at the University of Zurich, Jung had lectured there from 1905 to 1913. He had resigned "[c]onsciously, deliberately," feeling that he had to make a "choice of either continuing my academic career . . . or following the laws of my inner personality." It would have been "unfair to continue teaching young students when my own intellectual situation was nothing but a mass of doubts" (*Memories*, pp. 218–219).

[57] That is, a fixed point outside one's own sphere, offering "the possibility of objective measurement" (Jung, 1926 [1924], § 163). Jung repeatedly stressed that in psychology no such outside standpoint exists. For further references to this, see Jung & Schmid, 2013, pp. 15–16.

[58] Clark University, of course, being the university at which Jung and Freud had lectured in 1909, and received the degree of Doctor of Laws *honoris causa* (cf. Rosenzweig, 1992; Burnham, 2012). The series was edited by Carl Murchison and published by Clark University Press, Worcester, MA. The first traceable volume is from 1925.

not spoken to the younger generation for some twenty years. Consequently, I fear that I shall at times be off the mark. Should this occur, I would ask you to send me your questions through the post. But, please: within the scope of these lectures, rather than broaching the future of European currencies, for instance, or the prospects of National Socialism, etc.

I have called the psychology that I endeavor to discuss in these lectures "Modern Psychology." I have chosen such a general title, because the matters at hand are of a very general nature. Instead of engaging with specific doctrines, my aim is to paint a picture based on immediate experience in order to depict the development of modern psychological ideas.

Psychology did not suddenly spring into existence; one could say that it is as old as civilization itself. Obviously, psychology has always been with us, ever since human life, outstanding minds, personages, and psychological demonstrations have existed. In ancient times, there was the science of astrology, which has always appeared in the wake of culture all over the world. It is a kind of psychology, and alchemy is another unconscious form. This is an extremely peculiar form, however, a so-called projected psychology, in which the psyche is seen as entirely outside man, and is projected into the stars or into matter.[59]

But I do not intend at present to speak of those days. In this short introduction to "Modern Psychology," I shall take you back only to its first beginnings as a conscious science.[60] Psychology proper appears only with the dawn of the age of Enlightenment at the end of the seventeenth century, and we will follow its development through a long line of philosophers and scientists who made the manifestations of the psyche their field of study.

Still for Descartes (1596–1650),[61] the soul is quite simply thought directed by the will. In his time, the whole of scientific interest was not yet

---

[59] Cf. Jung, 1988 [1934–1939], p. 1496: "[O]ur whole mental life, our consciousness, began with projections . . . and it is interesting that those internal contents, which made the foundation of real consciousness, were projected the farthest into space—into the stars. So the first science was astrology."

[60] MS: *bewusste Wissenschaft*; that is, a psychology that is conscious, aware, of being a "psychology."

[61] René Descartes (1596–1650), the famous French philosopher and mathematician, most known for his dictum "cogito ergo sum," and his highly influential (and controversial) dualistic view of the mind–body problem (*res cogitans* vs. *res extensa*—mind is essentially thought, and body is essentially extension). His book *Meditationes de prima philosophia* (1641) is considered a classic contribution to Rationalism. In his theory, the soul is, in contrast to the body, an immaterial, unitary, and indestructible substance. It is always thinking, because thinking (*cogitatio*) is part of its essence. Thinking is guided by the will, which has to give its assent (*assensus*) to the judgment (*actus iudicandi*).

focused on the human soul, but flowed outward to concrete objects. The age of science coincided with the age of discovery, that is, the discovery of the surface of the world. Thus, science was only interested in what could be touched. The external world was thoroughly explored, but no one looked inward. While all kinds of psychic phenomena existed, of course, they fell into the domain of the dogmatic symbol. The soul was assumed to be known, and everything concerning it was left to the care of the Church. Phenomena of the soul occurred exclusively within the framework of the Church, in the form of religious, mystical, and metaphysical experiences, and were subject to the judgement of the priest. As long as this dogmatic symbol was a living thing, in which man felt contained, no psychological problems existed.

This strange fact—namely, that phenomena of the soul were still contained within the religious sphere—holds true wherever religion is still alive. There, the life of the soul finds valid expression in symbols, and what remains with the individual is in essence his consciousness, since everything else is already expressed in religious forms. For instance, a highly educated Catholic came up to me after a lecture, and remarked: "Dr. Jung, I am surprised that you go to such great pains with psychology, why you struggle with such problems; these are not problems, surely! Whenever doubt seizes me, I quite simply query my bishop, who might ask his cardinal, and eventually turn to Rome. After all, they must have gained more experience over 2,000 years than you have!"

For such people, psychological problems simply do not exist. This was the case for the whole of Europe deep into the first half of the nineteenth century, and this condition still remains undisturbed for those who feel secure in a living and effective religious form. In Buddhism, Islam, Confucianism, and so forth, too, the life of the soul is expressed in symbols.

Essentially, science rested not upon any fundamental doubt, but rather upon the doubt about the secondary manifestations of a truth already revealed. We must not overlook this fact. Thus, for instance, where people are still living within the framework of living symbols, our psychology lacks a point of attack altogether. For such people, these problems effectively do not exist. But once doubt sneaks in, the life in the symbol gutters out, and actual psychology begins.

As I mentioned, at the time when the great seafarers were discovering new continents, something freed itself, something which could no longer be contained in the dogmatic symbol. At first, one did not know what this was. It showed itself in a sudden longing for something from which the

Renaissance subsequently emerged. The Renaissance arose out of what, through doubt, had freed itself from Christianity. This was actually the first time that a psychological problem manifested itself.

Those of you who have read Jakob Burckhardt's study of the Renaissance might have stumbled over a small reference to a book entitled *Hypnerotomachia Poliphili*,[62] written by a monk, Francesco Colonna. The title means "sleep-love-conflict," that is to say it is highly symbolic. It was translated at the end of the sixteenth century by an otherwise unknown Frenchman as *Le songe de Poliphile*.[63]

The title refers to Polia,[64] or Madame Polia, the heroine of the conflict. The story begins with the hero—that is, the dreamer of a long dream—losing his way in the Black Forest, which the Italians considered an *ultima Thule*[65] at the time, and where unicorns were still said to roam. A wolf appears to him and leads him to the ruins of a sunken city with temples. Its architecture is that of the Renaissance—the whole of psychology was expressed in the form of architecture in the Renaissance. He steps into the dark entrance of one of the temples. After a while, he wishes to leave the temple again. He gets a somewhat uncanny feeling. But a great dragon appears in the doorway and blocks his way. In what follows, and since he can only go forward, he is compelled to experience everything that has happened to this sunken city. Through endless adventures, he is constantly looking for Madame Polia. Even though we do not know who this figure is, we can nonetheless venture a guess: Lady Soul. Eventually, he reaches the royal court. He is promised that he will be escorted to the Island of the Blessed where he will be wed to his beloved Polia. Upon

---

[62] Jacob Burckhardt (1818–1897), noted Swiss historian of art and culture, and one of the major progenitors of cultural history. His best known work is the one on the Renaissance, quoted here by Jung (Burckhardt, 1860, the reference on p. 186, Engl. ed.). Jung used the second German edition of 1869 (*Transformations*, § 21[23]). Regarding Burckhardt and the *Hypnerotomachia*, Jung remarked: "It is perhaps significant that this book, so important for the psychology of the Renaissance, was carefully avoided by the bachelor Jacob Buckhardt" (1963, § 1279[2]; my trans., only in *GW*, not in *CW*; see also note 67). As a cultural historian, however, Burckhardt was more interested in other, e.g., architectural, aspects of this book than in the psychology of the novel. Jung repeatedly quoted Burckhardt's notion of "primordial images" (e.g., 1917–1942, § 101; *Types*, definitions: image) in connection with his own of the "archetypes."

[63] Colonna, 1499. Béroalde de Verville's translation appeared in 1600; the first complete English version was published in 1999, five hundred years after the original (see Bibliography). In this book, Francesco Colonna describes his dream of an adventurous journey, in which he (as a monk) searches for the Lady Soul. The identity of Colonna is contended. He could have been a Venetian Dominican, or a Roman nobleman.

[64] φιλία (philia) = Greek for love; Poliphilus = the one who loves Polia.

[65] A mythical place beyond the borders of the known world.

arriving on the island, he hears a ringing and awakens. It is the morning of May 1st. *Hélas!*[66]

At the time, the story was said to be particularly profound and mysterious, and even thought to be a divine revelation. Later, it was considered to be so banal that Jacob Burckhardt did not even read it. Incidentally, the book is now a bibliographical rarity. Even the French edition has a collector's value of approximately five hundred Swiss francs. It took me great pains to read it at the time.[67]

The *Hypnerotomachia Poliphili* is an important *document humain*, and actually represents the secret psychology of the Renaissance, namely, that which had struggled free from the grip of the symbol. Significantly, its author was a monk, even though he expressed himself in a pagan way. Strictly speaking, he would have been obliged to express what moved his soul in Marianist terms, that is, through the symbol of the Mother of God, and yet he chose not to. His is an involuntary psychology, typical and in a way symptomatic of an entire historical period. It reveals what liberated itself at that time, and summons the world of the ancient Greek Gods to express this in one way or another. Under the cloak of this allegory, he describes the descent into the underworld of the psyche. Dame Polia held something for him that he could not find in the Madonna.

If this interpretation is correct, we must expect that anyone who became involved with this new symbol in subsequent centuries could no longer be a real Catholic. When we come to the philosophers, who took the path of psychological discovery and who became the founders of this comparatively modern science, we find that they were indeed almost without exception Protestants. In earlier days, the healing of the psyche was regarded as Christ's prerogative, the task belonged to religion, for we suffered then only as part of a collective suffering. It was a new point of view to look upon the individual psyche as something whole that also suffers individually. The Protestant is the natural seeker in the field of psychological research, for he no longer has a symbol in which he can express himself, and therefore his sense of incompleteness makes him uneasy; he searches, he is active and restless. He will set out to explore

[66] French, alas!

[67] In 1947, Linda Fierz-David (the wife of Hans Eduard Fierz, C. G. Jung's friend and professor of chemistry at the ETH) published a monograph on *The Dream of Poliphilo* (Engl. ed. 1950), to which Jung wrote a foreword, in which he told of his first encounter with the book: "I set about reading the book, but soon got lost in the mazes of its architectural fantasies, which no human being can enjoy today. Probably the same thing has happened to many a reader, and we can only sympathize with Jacob Burckhardt, who dismissed it with a brief mention while bothering little about its contents" (1947 [1946], § 1749).

every nook and cranny of the world in search of what he lacks, and he may have recourse to antiquity and learn about it, or will often reach out to other faiths, such as theosophy, Christian Science, Buddhism, etc., to find it there.

Eventually, he will come upon his soul and ask: Why is there something inside us that desires something else? "Why does my spiritual life no longer satisfy me?" is particularly the problem of the Protestant; he thinks that it should, but the fact remains that it does not, and that he is often troubled with neurotic symptoms. Thus, psychology was at first an entirely Protestant affair, then it became the business of the Enlightenment man, the skeptic, and the freethinker. For we can neither escape the fact that something rankles us nor that we are terribly nervous. Ultimately, psychology thus became a matter for the doctor. He must attend to those who have fallen into a profound doubt, and out of the symbol.

In what follows, I shall discuss in greater depth the development outlined so far. Specifically, I shall adduce a number of dates that will help us trace the gradual progress of psychology over the past centuries.

Gottfried Wilhelm Leibniz (1646–1716),[68] an encyclopedic genius and a celebrated philosopher in his day, made the first explicit contribution to what we call psychology today. I shall mention only a few key points here that were essential to the emergence of modern psychology. Very often, by the way, the teachings of the older philosophers are truths that then fell into oblivion for a long time.

Leibniz's central concept is what he called the *petites perceptions* [minute perceptions], *perceptions imperceptibles* [imperceptible perceptions], or *perceptions insensibles* [unfelt perceptions][69]: He thinks of perceptions as representations, since a perception is at the same time a representation.

[68] Gottfried Wilhelm Leibniz (1646–1716), German mathematician and philosopher, known as the last "universal genius." He made major contributions to the fields of metaphysics, epistemology, logic, and philosophy of religion, as well as mathematics (infinitesimal calculus), physics, geology, jurisprudence, and history. He is considered one of the great seventeenth-century advocates of Rationalism. Known for his theories of the monads and of pre-established harmony (to which Jung will refer in his writings on synchronicity; cf. Jung, 1952, §§ 927–928). Famous is his (often misunderstood) dictum of ours as "the best of all possible worlds."

[69] "At every moment there is in us an infinity of perceptions, unaccompanied by awareness or reflection; that is, of alterations in the soul itself, of which we are unaware because these impressions are either too minute and too numerous, or else too unvarying, so that they are not sufficiently distinctive on their own" (Leibniz, 1981 [1704–1706], p. 53). The infinity of petites perceptions is, so to speak, epistemological white noise.

Leibniz cites as an example the experiment involving blue and yellow powder.[70] When they are mixed insufficiently, blue and yellow grains of powder are distinctly perceptible. But when they are mixed thoroughly, only green powder is perceptible, even though the powder still consists of blue and yellow grains. While it looks green, it is in reality yellow and blue. We perceive these two colors—blue and yellow—unconsciously, that is to say, beneath the threshold. They are imperceptible. Leibniz tried to find a psychological meaning to his experiments and sought to make analogies to similar processes that take place in the human mind: something happens in me of which I am not aware. Here we first chance upon the conception of a soul that is not conscious. Descartes still considered the soul to be nothing other than thought.

For Leibniz, these "minute perceptions" contrast with another psychological principle: the principle of the intellect or the idea. Ideas and innate truths do not exist as actualities in us, however, but instead as some kind of dispositions that experience must fill out in order for them to become perceptible: "*c'est ainsi que les idées et les vérités nous sont innées comme des inclinations, des dispositons, des habitudes ou des virtualités.*"[71] It is like a drawing that, although it has already been made, is invisible, but nonetheless exists, because when we douse it with powder it suddenly becomes visible.

Perceptions are the opportunities for and the causes of rendering conscious innate ideas and dispositions. Leibniz thus anticipated the idea of innate dispositions, that is, images in which we accumulate and shape experience. For him, representations are a kind of powder that is spread over the inborn or unconscious ideas. These ideas, which came already very close to modern psychology, remained latent for a very long time, as is often the case with ideas when the time is not yet ripe for them.

His younger contemporary Christian August Wolff (1679–1754)[72] initiated another line of thinking. Wolff limited his discussion entirely

---

[70] "[W]hen we perceive the colour green in a mixture of yellow and blue powder, we sense only yellow and blue finely mixed, even though we do not notice this, but rather fashion some new thing for ourselves [*novum aliquod ens ex nobis fingentes*]" (Leibniz, 1684, p. 426).

[71] Ideas and truths are innate inside us as "inclinations, dispositions, tendencies or natural potentialities and not as actualities" (Leibniz, 1981 [1704–1706], p. 52).

[72] Christian Wolff (1679–1754), arguably the most important German philosopher in the early and middle portion of the eighteenth century, between Leibniz (with whom he was acquainted and corresponded [Leibniz & Wolff, 1860]) and Kant. He wanted to base theological truths on mathematical evidence, his philosophy being a systematic development of Rationalism. Accused of atheism, he was ousted in 1723 from his first chair in Halle, and ordered by the king to leave Prussia within 48 hours or be hanged, causing one of the most

to consciousness, and divided his psychology into two parts: firstly, empirical psychology, which considers in particular the cognitive faculty and the activity of consciousness; and secondly, rational or speculative psychology, which centers on desire and the interrelations between body and soul.[73]

Wolff considered the "soul" a simple substance, endowed with three powers: the representative faculty, the appetitive faculty, and the cognitive or cogniscitive faculty.[74] However, he considers thinking to be the essence of the soul.[75] In Wolff, we encounter for the first time the notion that psychology could be experience and that one could even experiment with it, which was a completely new idea. Wolff's psychology is the first ever experiential psychology.[76]

Johann Nikolaus Tetens (1736–1807)[77] went even a step further. He is the actual founder of experimental, physiological psychology, which later

celebrated academic dramas in the eighteenth century. He had a wide following of "Wolffians," making him the founder of the first German philosophical "school," dominating Germany until the rise of Kantianism. Interestingly, in connection with Jung, his preoccupation with Confucius, and Chinese philosophy (cf. his famous lecture "On the Practical Philosophy of the Chinese" [1721]), is considered an early highlight of the encounter between Western and Eastern philosophy. His complete writings have been published since 1962 in an annotated edition (Wolff, 1962 sqq.).

[73] Wolff defined psychology as that "part of philosophy that deals with the soul" [*pars philosophiae, quae de anime agit*] (Wolff, 1728, § 58, p. 29). He then distinguished between *psychologia empirica* and *psychologia rationalis*. In the latter, "we derive, solely from the concept of the human soul, a priori everything that can be seen as belonging to it a posteriori, and also that which is deduced from observations [of the soul]" [*In Psychologia rationali ex unico animae humanae conceptu derivamus a priori omnia, quae eidem competere a posteriori observantur & ex quibus observatis deducuntur*] (ibid., § 112, p. 151).

[74] Wolff was a representative of "faculty psychology" [*Vermögenspsychologie*], a point of view that conceived the human mind as consisting of separate powers or faculties, which was a widespread concept during much of the nineteenth century.

[75] *Cogitatio igitur est actus animae, quo sibi sui rerumque aliarum extra se conscia est* [Thinking is therefore the act of the soul by which it becomes conscious of itself and of the other things outside of itself] (1732, § 23), quoted by Jung in *Transformations*, ed. 1991, p. 25. On Wolff's general views on the soul, cf. Wolff, 1719–1720, 1733.

[76] "Practical philosophy is of the utmost importance, and that is why it is so important that we do not proceed from principles that could be doubted. We can only base the truths of practical philosophy, therefore, on basic principles that are obviously supported by experience in psychology" [*Philosophia practica est maximi momenti; quae igitur maximi sunt momenti, istiusmodi principiis superstruere noluimus, quae in disceptationem vocantur. Ea de causa veritates philosophiae practicae non superstruimus nisi principiis, quae per experientiam in Psychologia evidenter stabiliunter*] (Wolff, 1728, p. 52). On Wolff as a pioneer of psychology as a natural science, see also Jung, 1946b, § 345.

[77] Johannes Nikolaus Tetens (1736–1807), German philosopher, mathematician, and scientist of the Enlightenment. In the wake of Christian Wolff, who himself drew on John Locke, Tetens drew on English Empiricism. In English-speaking countries, he has been called "the German Hume," having studied and popularized Hume's work in the German-speaking

flourished before World War One in the era of Wilhelm Wundt (1832–1920).[78] Tetens was influenced by the English physiological approach to psychology, as represented by David Hartley (1705–1757).[79] Tetens was the first to measure the sensations of light, hearing, and touch. He espoused a wholly empirical approach and did not consider doctrines to be eternal truths, but, rather as did the English, to be mere "working hypotheses."

This age peaked in the great critical era whose pre-eminent figure was Immanuel Kant (1724–1804).[80] His critique of knowledge also imposed

---

world. His main work, *Philosophische Versuche über die menschliche Natur und ihre Entwickelung* [Philosophical essays on human nature and its development] (1777), sought to combine Hume's empiricism (cf. Hume, 1739–1740) with the philosophy of Leibniz and Wolff. He tried to make a psychological analysis of the soul with the methods of natural science. His work also was important for Immanuel Kant, whom he is supposed to have introduced to phenomenological thought and to the empiricism/transcendence dualism.

[78] Wilhelm Wundt (1832–1920), German physician, psychologist, physiologist, and philosopher, generally considered the "father" of psychology as a separate natural science in general, and of experimental psychology in particular (although according to Jung the credit actually goes to Tetens). Founder of the first psychological laboratory in Europe (1879) and the first journal for psychological research (1881). Wundt played a central role in the nascent field of psychology, not least on Freud (through his ethno-psychological writings) and Jung (association experiments). His legacy in psychology today, however, is a subject of continuing debate.

[79] David Hartley (1705–1757), English philosopher, scientist, and mystic, also a practicing physician and vegetarian. His central concept of "association" led to the school of "association psychology" in the nineteenth century (James Mill, John Stewart Mill, William B. Carpenter, Alexander Bain). His principal work, *Observations on Man, His Frame, His Duty, and His Expectations* (1749), studied humans as physical beings (frame), psychological and moral beings (duty), and as religious beings (expectations), representing a wide-ranging synthesis of neurology, moral psychology, and spirituality. His "physiological approach to psychology" was to start with "corporeal causes"—neurological processes ("vibrations" in the brain)—and then to ask how such processes generated consciousness, perceptions, thoughts, etc. He affirmed the unity of body and mind, and trusted in universal salvation and the eventual overcoming of the chasm between hell and heaven. His theories gave rise to heated controversies at the time, but were also strongly supported by influential figures such as Joseph Priestley. On Hartley and Priestley see also Lecture 2, and note 101).

[80] Immanuel Kant (1724–1804), from Königsberg/Kaliningrad (then in East Prussia/Germany, now in Russia), the central figure in modern philosophy. Jung's interest in Kant dates back to his adolescence when, studying and admiring Schopenhauer, he "became increasingly impressed by [the latter's] relation to Kant" (*Memories*, pp. 88–89), and he started to study the *Critique of Pure Reason* (cf. also his *Zofingia Lectures* (1983 [2000]), which he found "an even greater illumination than Schopenhauer's work. To a student at the Jung Institute in the 1950's, Jung exclaimed, 'Kant is my philosopher,' and Kant's critique formed the basis for his understanding of the boundaries of knowledge" (Shamdasani, 2012, p. 22).—The 1780s, when Kant published *The Critique of Pure Reason* (1781/1787), are now considered a transitional decade—what Jung called "the great critical era"—in which the Enlightenment was already in a state of crisis, and the cultural

boundaries on psychology. In particular, Kant contested its possibility of being a science, arguing instead that it was at best a "discipline." Despite his skepticism Kant was not opposed to psychology, but actually took a profound interest in it. His views on the subject are somewhat contradictory and awkward, however, and are consequently discarded by "true" Kantians.[81] In his *Anthropology* he follows Leibniz's thinking, and speaks of "obscure representations," that is to say, representations that we have without being conscious of them.[82]

---

balance shifted toward Romanticism. See also Lecture 2, where Kant's views are treated in more detail.

[81] In *Metaphysical Foundations of Natural Science* (1786), Kant declared that "empirical psychology [must] be removed from the rank of what may be termed a natural science proper; firstly, because mathematics is inapplicable to the phenomena of the internal sense and its laws. . . . [Secondly,] because in it the manifold of internal observation is only separated in thought, but cannot be kept separate and be connected again at pleasure; still less is another thinking subject amenable to investigations of this kind, and even the observation itself alters and distorts the state of the object observed. It can never therefore be anything more than . . . a natural description of the soul, but not a science of the soul, nor even a psychological experimental doctrine" (Preface). His own *Anthropology from a Pragmatic Point of View* (1798), however, is itself in large part an empirical psychology. For a contemporary assessment of Kant on psychology, see, e.g., Sturm, 2001.

[82] In *Anthropology*, he defines obscure representations as "sensuous intuitions and sensations of which we are not conscious, even though we can undoubtedly conclude that we have them" (Kant, 1798, Engl. ed. 2006, p. 24). Similarly, he states in his lectures on metaphysics: "Our representations are either obscure or clear, etc. Obscure representations are those of which I am not immediately conscious, but nevertheless can become conscious through inferences" (Kant, 1902/1910 sqq., vol. 29, p. 879).

Lecture 2

27 OCTOBER 1933

SUBMITTED QUESTIONS

The first question is about Leibniz's *perceptions insensibles*[83] and asks for a concrete psychological equivalent to Leibniz's experiment with the blue and yellow powder. Seen from some distance, the blue and yellow powder appears to be green.

Our daily life abounds in concrete psychological examples of Leibniz's "unconscious perceptions" as illustrated by the above experiment. These are the many things we do unconsciously. We look, for instance, at our watch, but we have to consult it again if asked the time a minute later, yet we perceived it unconsciously. There are other cases, such as riding a bicycle, where the process is almost wholly unconscious and if, while actually riding a bicycle, we suddenly become aware of the unconscious perceptions by which we keep our balance, it may prove actually dangerous. These *petites perceptions* become visible and invisible in a manner analogous to the blue and yellow particles in the green powder.

The second question concerns Catholicism. The correspondent has written to inquire whether I would argue that Catholicism is not a psychology. "Why did you argue in your last lecture that psychology is so modern?"

Healing used to be considered the prerogative of Christ. It was a religious matter. In this sense, it is new to regard the human psyche as a whole and to experience ourselves as a suffering totality. I have never disputed the fact, however, that people possessed a soul already in those times. Psychic problems have always existed, but people were not yet ready to develop a science about this. Thus, religions were the method

[83] At the beginning of the first question, the lecture notes read: "Leibniz: dispersion insensible." Very probably, this is a hearing mistake for "*die* [the] *perceptions insensibles*," because the question is precisely about that.

applied to overcome these difficulties. Jesus was a "savior" after all, a doctor. He could heal illnesses. The sufferings of the soul were supposed to be healed in this way. If total faith is possible within a religious system, then adequate healing of the suffering soul, too, can occur within this system. Nervous disorders arise from disturbances of the soul life—and not, for instance, from poor sleep or eating too many potatoes. Thus, an infantry captain suffering from foot pain once came to see me.[84] In the case of a roofer, it would have been vertigo; in the scholar's case, perhaps an affliction of the eye. Body and soul are one being. We do not suffer in just an isolated corner of the whole. If a part suffers, the whole suffers. So the question was why no psychology yet existed in the Middle Ages. This is a misuse of the term "psychology": people sometimes speak of "Mr. Jones's psychology" when they actually mean the psyche of Mr. Jones.[85]

***

Now, let us return to the topic of today's lecture. As we have heard, the age of the critique of knowledge began towards the end of the eighteenth century, with Immanuel Kant (1724–1804) as its leading figure. His [concept of] "obscure representations"[86] follows Leibniz's train of thought and carries his ideas further. In the first volume of his *Anthropology*, he speaks of "representations that we have without being conscious of them."[87] We could also refer to perceptions. "Representations," however, is the more general term. The famous statement that there is nothing in the mind that was not previously in the sensations[88] has in fact not been proven.

Kant adduces the following example: From afar, he sees a person standing in a meadow. What he sees, however, is in actual fact merely a shadow, because he cannot discern any details—limbs, eyes, nose, and so forth. Yet he nonetheless has the idea or representation that the figure is a person. This is essentially Leibniz's idea. Kant's conclusion, however, reaches far beyond the latter's, and it affects the field of psychology much more profoundly—namely, that "only a few places on the vast map of

[84] Jung was an officer in the Reserve in the Swiss army, and was—like every able-bodied male Swiss citizen—drafted once a year for military service, in his case, as an army doctor.

[85] In fact, throughout his work—and also in these lectures—Jung often himself uses "psychology" for both the psyche, or mental make-up, of somebody, and the theory or science of the psyche.

[86] See the previous lecture and note 82.

[87] Ibid., p. 23.

[88] See below, **p. 15** and note 96.

our mind are illuminated," and "the field of *obscure* representations is the largest in the human being":

> The field of sensuous intuitions and sensations of which we are not conscious, even though we can undoubtedly conclude that we have them; that is, *obscure* representations in the human being (and thus also in animals), is immense. Clear representations, on the other hand, contain only infinitely few points of this field which lie open to consciousness; so that as it were only a few places on the *vast* map of our mind are *illuminated*. This can inspire us with wonder over our own being, for a higher being need only call "Let there be light!" and then, without the slightest cooperation on our part . . . as it were set half a world before his eyes. . . . Thus the field of *obscure* representations is the largest in the human being.[89]

These reflections cast the subject matter of psychology in an entirely different light and delineate it more sharply. For it could be argued that all psyches are individual psyches, that no such thing as a collective psyche exists, and that the psyche is nothing other than consciousness, as Professor Krüger claimed at the last Congress of Psychology.[90] Consciousness is after all an individual phenomenon. But if you ask a primitive whether he has an individual psyche, whether he is distinct from his fellow man or from his surroundings, he will not be at all certain. When you are among primitives you hardly dare to kill a crocodile, for the primitive says: "I am also that crocodile." Kant thus initially assumes that our consciousness, that is to say, that which we are clearly conscious of, corresponds solely to some few illuminated points, and everything else lies in darkness. Just as a scholar may remark: "If I knew everything that I have already forgotten, I would be the most learned of all men." The circumference of consciousness is thus very limited, and that of the unconscious is consequently all the greater. Kant was thus the first to recognize this fundamental truth in psychology.

After Kant, the epoch of empirical psychology in Germany came to a temporary end. There followed a period of great metaphysical speculation,

[89] Ibid., pp. 24–25.

[90] Felix Krüger (1874–1948), professor of psychology at the University of Leipzig. Krüger, who had been a member of a "National-Socialist combat group for German culture" since 1930, was appointed Rector of the University of Leipzig in 1935. In 1937, however, he was suspended on account of his Jewish ancestry. He retired in 1938, and emigrated to Switzerland in 1944.—Jung is probably referring to the Thirteenth Congress of the German Psychological Society in Leipzig of October 16–19, 1933, that is, just before Jung began his lectures. Cf. Klemm, 1934.

in which the principle of imagination or fantasy reacted against the critique of pure reason. Hegel and Schelling were in reality metaphysical speculators, but when you examine their writings carefully—particularly those of Hegel—you will see that they are full of projected psychology.

Georg Wilhelm Hegel (1770–1831)[91] allowed fantasy to become outrageously speculative. There is no doubt at all that today Hegel would not be a philosopher, but a psychologist. He himself was not aware of this, however, and referred to his work as "philosophy."[92] But in effect it constituted a psychology of the unconscious. Essentially, he deduced a psychology of the dark field, and in certain theses he actually talked about the psychology of the unconscious.

Likewise, Schelling (1775–1854)[93] espoused a positive stance toward the unconscious, whereas for Kant this remained a negative boundary concept. Schelling maintained that this "eternally Unconscious, which, as were it the eternal sun in the kingdom of spirits, is hidden by its own untroubled light," is the absolute ground of consciousness. The unconscious thus becomes the primordial motherly foundation. This field is not some shadowy Hades but, as Schelling states, the sun from which consciousness emerges, being hardly more than a reflection. He adds:

> and although itself never becoming Object, [it] impresses its identity on all free actions, is withal the same for all intelligences, the invisible root of which all intelligences are only the powers, and the eternal mediator between the self-determining subjective in us and the objective or intuited, at once the ground of conformity to law in freedom, and of freedom in conformity to law.[94]

---

[91] Georg Wilhelm Friedrich Hegel (1770–1831), the famous German philosopher and pre-eminent representative of German (philosophical) Idealism.

[92] A view that Jung expressed also elsewhere: "Hegel, that great psychologist in philosopher's garb" (Jung, 1935a, § 1734); or "Hegel was a psychologist in disguise who projected great truths out of the subjective sphere into a cosmos he himself had created" (Jung, 1946b, § 358). To Friedrich Seifert he wrote in 1935: "I have always been of the opinion that Hegel is a psychologist in disguise, just like I am a philosopher in disguise" (1973, p. 194).—In *Memories*, Jung says that when he first encountered his writings in his adolescence, "Hegel put me off by his language, as arrogant as it was laborious; I regarded him with downright mistrust. He seemed to me like a man who was caged in the edifice of his own words and was pompously gesticulating in his prison" (p. 87).

[93] Friedrich Wilhelm Joseph Ritter von Schelling (1775–1854), another main representative of German Idealism. His philosophy is situated between Fichte, his mentor prior to 1800, and Hegel, his former university roommate and erstwhile friend.

[94] *Dieses ewig Unbewußte, was, gleichsam die ewige Sonne im Reich der Geister, durch sein eigenes ungetrübtes Licht sich verbirgt, und obgleich es nie Object wird, doch allen freyen Handlungen seine Identität aufdrückt, ist zugleich dasselbe für alle Intelligenzen, die unsichtbare Wurzel, wovon alle Intelligenzen nur die Potenzen sind, und das ewig*

Schelling thus places complete emphasis on the unconscious. I would like to draw your attention particularly to the passage, "it is the same for all intelligences": whereas intelligences are single, the ultimate basis is the same everywhere. The ultimate basis is not differential, but universal. Major philosophical systems later emerged from this very idea of Schelling's.

While empirical psychology was in a sorry state in Germany for the time being, it rose to prominence in England where modern science espoused it as an important mode of thinking from early on, and particularly after Kant.

George Berkeley (1685–1753)[95] is the first English empirical psychologist. As an empiricist, he made sensations his starting point, like Christian Wolff. When one neither sees, hears, nor feels anything, one assumes that there is nothing in the mind either, true to the Latin saying: *Nihil est in intellectu, quod non antea fuerit in sensu.*[96] Berkeley realized, however, that sensations do not remain disparate, but coalesce into a whole, and discovered the perception of one's own senses as a factor equal to the object perceived. Out of this fusion of subject and object Berkeley constructed the psychological concept of space.

David Hume (1711–1776)[97] argued along similar lines. He considered for the first time the relation between representations and sensations, and

---

Vermittelnde des sich selbst bestimmenden Subjectiven in uns, und des Objectiven, oder Anschauenden, zugleich der Grund der Gesetzmäßigkeit in der Freyheit, und der Freyheit in der Gesetzmäßigkeit des Objectiven (Schelling, 1800, p. 434; also in *Sämmtliche Werke*, I, 3, p. 600). Jung had already referred to this passage in a lecture at the ETH about two years earlier: "In Schelling the 'eternally unconscious' is the absolute ground of consciousness" (Jung, 1932, § 1223).

[95] George Berkeley (1695–1753), or Bishop Berkeley, Anglo-Irish philosopher, known for his radicalization of Locke's sensualism and his immaterialism, as summarized in his notion that "to be is to be perceived" [*esse est percipi*]: "For as to what is said of the absolute existence of unthinking things without any relation to their being perceived, that is to me perfectly unintelligible. Their *esse* is *percipi*, nor is it possible they should have any existence out of the minds or thinking things which perceive them" (Berkeley, 1710, part 1, § 3). His philosophy is at the basis of the well-known thought experiment: "If a tree falls in a forest and no one is around to hear it, does it make a sound?" Incidentally, the city of Berkeley, California, was named after him, while the pronunciation was Americanized.

[96] Latin, there is nothing in the intellect (mind) that was not previously in the sense (the sensations). This maxim of the philosophical doctrine of sensualism is found in various slightly different versions (e.g., in Thomas Aquinas, 1256–59, quaestio 2, articulus 3, arg. 19; or, most famously, in John Locke). Jung uses the wording quoted by Schopenhauer (1819 [1887], p. 258), who attributed this judgement to Aristotle.

[97] David Hume (1711–1776), very influential Scottish philosopher, historian, and economist; like John Locke and George Berkeley a main representative of British Empiricism. Kant, for instance, credits him with being the one who "many years ago first interrupted my dogmatic slumber, and gave my investigations in the field of speculative philosophy quite a new direction" (1783 [1902], p. 7).

derived the former from the latter. He espoused Berkeley's idea of coalescence for his representations and enquired into the law of coalescence. He argued that their association occurs on account of similarity, coexistence in time and space, and causality.[98] Association occurs by means of a "gentle force,"[99] similar to the law of gravity and its effect on heavenly bodies. Representations thus reciprocally attract each other.

David Hartley (1705–1757),[100] one of Hume's contemporaries, ventured to explore more complex psychic phenomena on this basis. He applied this fusion principle to the highest complexities of the mind, which he also explained in terms of the fusion into a whole of rapidly recurring or simultaneous sensations.

In the work of Joseph Priestley (1733–1804),[101] this mechanistic attempt shifted towards materialism through his identification of these psychic processes with processes in the brain. Such identification obviously had consequences for psychology.

The concept of instinct, or so-called common sense, appeared with the Scottish School of philosophy, whose leading exponents included Thomas Reid (1710–1796)[102] and William Hamilton (1788–1856).[103]

---

[98] "It is evident that there is a principle of connexion between the different thoughts or ideas of the mind, and that in their appearance to the memory or imagination, they introduce each other with a certain degree of method and regularity. . . . To me, there appear to be only three principles of connexion among ideas, namely, Resemblance, Contiguity in time or place, and Cause or Effect" (Hume, 1748 [1993], Section III, p. 14; ital. in orig.).

[99] We should regard the three principles of association "as a gentle force, which commonly prevails" (Hume, 1739–1740, Book I, Part 1, Section 4).

[100] See Lecture 1 and note 79.

[101] Joseph Priestley (1733–1804), English polymath—theologian, Dissenting (Unitarian) clergyman, natural philosopher, chemist, educator, and political theorist. He was the foremost British scientist of his age (for example, he is usually credited with the discovery of oxygen and other gases), although he himself viewed his scientific work as only secondary to his theological work. He published over 150 books on a wide range of topics, on science, philosophy, theology, history, grammar and language, politics, education, etc. Priestley supported David Hartley, and claimed that the book that influenced him the most, save the Bible, was the latter's *Observations on Man* (1749). Priestley abandoned dualism and considered perception and other mental powers to be properties of matter. His materialism involved assimilating the attributes of mind to matter as a natural extension of Hartley's correlation of mental associations with vibrations in the brain.

[102] Thomas Reid (1710–1796) was the founder of the Scottish School of Common Sense, and played an important role in the Scottish Enlightenment; critic of John Locke, George Berkeley, and David Hume. For Reid, common sense or *sensus communis*—the opinions of "the vulgar," those tenets that we cannot help but believe, given that we are constructed the way we are constructed—are at the foundation of all philosophical inquiry and knowledge.

[103] William Hamilton (1788–1856), Scottish philosopher of enormous erudition; although largely forgotten today, he was at his time regarded as a major intellectual figure of international importance. After visits to Germany, he brought German philosophy, above

Reid defined "common sense" as "that upon which all agree." For him, common sense is the indubitable source of knowledge, through which we also become acquainted with complex psychic processes. Consequently, psychology could limit itself in essence to simply describing what is established by common sense. The idea of looking at everything simply and objectively may sound frightfully banal at first sight, but this is the empirical point of view par excellence, and can only be reached by a complete sacrifice of judgements and opinions. So this way of looking at things is an invaluable contribution to psychology.

This is Rudyard Kipling's attitude in his *Just So Stories*, a book comprising a selection of outright silly stories.[104] It is in its place when applied to the fearful complexities of the human psyche. If we are able to say "it is just so," there is nothing to be done about it, and then we must confine ourselves to simply have a look at the whole of psychology and forget about our previous judgement. This involves a self-sacrifice as well as limiting ourselves to the objective. You will have the right attitude to psychology in general and to the difficult things that you will hear in the course of these lectures, if you can treat them as a "Just So Stories," as mere descriptions. Thus, English "common sense" is very warranted.

Dugald Stewart (1753–1828),[105] Reid's student and follower, was convinced that this descriptive method could turn psychology into a natural science, through its objective description of psychic processes, by the sacrifice of all opinions and by making no foregone conclusions. And this is a very good idea, because once we are able to separate a theory from an opinion, it begins to become a science. Confusion abounds today, because everyone believes that psychology is identical with his opinion about it. But this is just his own psyche, and by no means psychology!

Stewart postulated two laws of association, and this discrimination is important for pathology. First, there are voluntary, arbitrary associations arising from the active interference of consciousness. Second, there are involuntary, spontaneous, and simple associations that follow certain *a*

all the work of Kant, to the British Isles. He was an enthusiastic exponent of Reid, whose collected works he edited and annotated (1846). In 1854/1855 he also published a re-edition of the works of Dugald Stewart (see below.)

[104] Kipling, 1902. Rudyard Kipling (1865–1936), the very popular English writer and poet, now mostly known for his tales for children (e.g., the *Jungle Books* [1894, 1895]), but also (in)famous for his celebration of British imperialism. Nobel Prize in Literature, 1907.

[105] Dugald Stewart (1753–1828), Scottish Enlightenment philosopher and mathematician. He upheld Reid's psychological method and expounded the "common-sense" doctrine, but was also an original and influential philosopher of his own, being co-responsible for making the "Scottish philosophy" predominant in early nineteenth-century Europe.

priori laws, such as resemblance, contrariety, and vicinity in place, or vicinity in time.[106] Some processes of the psyche obey the will, others do not, but follow laws of their own. People incline to identify with one of these views, but both are equally true. Great truths, such as the existence of voluntary and involuntary actions, are lost time and again, only to be rediscovered over and over again.

---

[106] For the latter associations Stewart listed "the relations of Resemblance and Analogy, of Contrariety, and Vicinity in time and place, and those which arise from accidental coincidence in the sound of different words. These, in general, connect our thoughts together, when they are suffered to take their natural course." For the former associations, there "are the relations of Cause and Effect, of Means and End, of Premises and Conclusion; and those others, which regulate the train of thought in the mind of the Philosopher" (1792, pp. 213–214).

Lecture 3

3 NOVEMBER 1933

THE SEQUENCE OF the development of psychology that we have been following took us to the British Isles last time. Today we will turn to France, where the first psychologists appear at the rise of the Enlightenment in the early eighteenth century. This was the age of the Encyclopedists; knowledge was being accumulated, and the ideas of philosophers, such as Voltaire and Diderot, were being spread abroad. France was a very Catholic country at the time, and when such a country is being enlightened, then it is thoroughly enlightened indeed, that is to say, matters move from one extreme to another.

The first psychologist whom we encounter in France is probably Julien Offray de La Mettrie (1709–1759),[107] a physician and an outstanding man of his time. In 1748, Frederick the Great called him to Berlin,[108] where he died eleven years later. In one sense, his mind was quite modern; he was a true materialist and empiricist. His chief claim was that all life arises from dead matter, that the organic springs from the inorganic. The soul is thus an adjunct of the organic as it were. The discovery of the relations between the psyche and the brain bears fruit in that the psychic becomes dependent on the brain. La Mettrie says: "The brain has muscles to think, just as the legs have muscles to walk."[109] La Mettrie conceives

[107] Julien Offray de La Mettrie (1709–1751), French physician and philosopher; a radical, early materialist of the Enlightenment.

[108] In fact, La Mettrie's materialistic, atheistic, and anticlerical principles—which made him a pariah even among his fellow philosophers of the Enlightenment—caused such an outrage that he had to exile himself to the Netherlands. But even in that relatively tolerant state, his views, and the publication of his book, *L'homme machine* (1748), raised such a storm of protest that he was forced to leave again. (Even the publisher of this book felt compelled to preface it with a statement, in which he tried to justify publishing such a *livre hardi* [impertinent book].) In Berlin, La Mettrie enjoyed the tolerance of Frederick II (the "Great"), was allowed to practise as a physician, and was even appointed court reader.

[109] *Le Cerveau a ses muscles pour penser, comme les jambes pour marcher* (La Mettrie, 1748, pp. 77–78).

the living being as a machine, a machine that he likened to a clock mechanism consisting of various mainsprings.[110] In his famous book, *L'Homme Machine* (1748), he unabashedly holds the view that the soul is nothing other than a sensitive, material part of the brain.[111] This is a view that has persisted to this day.

Etienne Bonnot de Condillac (1715–1780),[112] who took holy orders and later became an abbé, was La Mettrie's contemporary, but survived him by many years. From his love affair with one Mademoiselle Ferrand, Condillac learned that all psychic life originates in sensation. He was ingenious enough to develop this insight into a philosophy. His principal work, *Traité des Sensations*, first appeared in 1754.[113] It was reprinted as late as. . . . [114] Significantly, it was not translated into German until 1870,[115] that is to say, until the golden age of materialism. Contrary to the general belief at that time that certain ideas are *a priori* innate in man, Condillac argued that the whole of the soul is empty. The mind would be an absolute *tabula rasa*. If we lived in isolation and spoke to no one, then nothing at all would come into being.

In working out their theories, philosophers have often sought a *point de repère*, that is, a point of reference, an idea, a metaphor, or even a material object, on which to develop them. Of Kant it is said that he had a listener who faithfully followed his lectures, year after year, always occupying the same place in the auditorium. In order to aid his concentration, and as a *point de repère*, Kant used to focus his vision on the top button

[110] *Le corps n'est qu'une horloge* [The body is nothing but a clock]; and: *Je ne me trompe point, le corps humain est une horloge, mais immense, & construite avec . . . d'Artifice & d'Habilité* [I am definitely not mistaken, the human body is a clock, but an immense one, artfully and ably constructed] (ibid., pp. 85, 93).

[111] *L'Ame . . . existe, & il a son siège dans le cerveau à l'origine des nerfs, par lesquels il exerce son empire, sur tout le reste du corps. Par là s'explique tout ce qui peut s'expliquer, jusqu'aux effets surprenants des maladies de l'Imagination* [The soul exists, and it has its seat in the brain at the origin of the nerves, through which it exercises its reign on the whole rest of the body. This explains everything that can be explained, even the surprising effects of the diseases of the imagination] (ibid., p. 78).

[112] Étienne Bonnot de Condillac (1715–1780), French philosopher, epistemologist, and economist. He systematically established the principles of Locke in France. He was an advocate of sensualism, or empirical sensationism, holding that all human faculty and knowledge are transformed sensation only.

[113] There Condillac expressly states: *Ce traité n'est . . . que le résultat des conversations que j'ai eues avec elle [i.e., Mlle. Ferrand]* [This treatise is nothing but the result of the conversations I had with Ms. Ferrand] (1754 [1798], p. 53).

[114] Here the lecture notes differ. Hannah has 1885, Kluger-Schärf 1785, Schmid both 1785 and 1885, Sidler 1794 and 1885. In any case, the book appeared in at least three reprints in 1788, 1792 and 1798 (see Bibliography).

[115] A first German translation appeared as early as in 1791, however (see Bibliography).

of this gentleman's waistcoat. And on one occasion, when the young man did not appear, the great philosopher found himself unable to proceed with his lecture![116]

Condillac's *point de repère* was his image of a man who is in fact not a real man, but a large dead statue that was endowed, however, with a sensory faculty.[117] Gradually, all the statue's senses awaken, first of all its sense of smell. Condillac sought to reconstruct the whole of thc human psyche from this statue and its sensations, without the help of any other hypotheses. This procedure is characteristic of the psychological attitude of this researcher. He was anxious to annihilate the extremely living, intangible, iridescent quality of the human soul, this absolutely ungraspable being, to turn it into stone, that is to say, into a kind of specimen, and to then probe it at a single, specific point. He kills the psychic matter, as it were, in order to able to dissect and study it.[118]

Occasionally, this is how reason deals with the psyche—namely, by killing it in order to bring it to life, as it were. This is the result of a particular mental attitude, which prevailed until the end of the nineteenth century. Condillac considered any possible content of the soul to be a "transformed sensation" [*sensation transformée*].[119] Here, a little bit of metaphysics enters into the equation because, according to Condillac, the soul is a sentient substance, but at the same time an immaterial substance—though this is difficult to imagine—that is to say, a subjectless sensation that wanders about aimlessly in the universe.

Perhaps you have heard of Rudolf Steiner.[120] In one of his books, he writes that before the planet Earth came into existence, a whole array of worlds existed, including gas-filled ones inhabited entirely by ethe-

---

[116] "He usually chose a listener who sat near him and read in his face whether he was being understood or not. Once he had begun to unfold his thought, the slightest disturbance in the auditorium could interrupt his train of thought. He was once distracted by a student who sat right in front of him with a button missing on his coat" (Gulyga, 1987, p. 79).

[117] Condillac imagines a statue organized inwardly like a man, animated by a soul that has never received an idea, into which no sense-impression has yet penetrated. He then unlocks its senses one by one.

[118] In his inaugural lecture at the ETH on 5 May 1934 (that is, a few months after the start of his regular lectures), Jung again referred to Condillac, criticizing him for his view that it would be "possible to investigate *isolated* psychic processes. There are no isolated psychic processes, just as there are no isolated life-processes" (1934a, § 197).

[119] For example: *Le desir . . . , les passions, l'amour, la haine, l'espérance, la crainte, la volonté. Tout cela n'est donc encore que la sensation transformée* [Desire, the passions, love, hate, hope, anxiety, the will. All this is also nothing but transformed sensation] (Condillac, 1754 [1798], p. 21).

[120] Rudolf Steiner (1861–1925), the well-known and influential Austrian esotericist and philosopher, founder of anthroposophy. On Jung and Steiner, see Wehr, 1972.

real beings. He observed further that the existence of sensations of taste has also been substantiated in this gas-filled sphere.[121] I refer here also to Christian Morgenstern's poem "The Knee," wherein "On earth there roams a lonely knee. It's just a knee, that's all."[122]

It is worth noting that the absoluteness of eighteenth-century French psychology rests firmly upon the Latin tradition. One early example for Condillac was Arnobius Africanus (350 CE),[123] a Latin Church father, who argued that the human soul is empty and of a material nature. Everything that enters it is based on sensory experience. His belief, which is shared by Christianity in general, is that the soul is either non-existent prior to baptism and enters the body over time or, should it exist, then solely in an utterly deplorable condition, that of original sin, necessitating divine enlightenment. The human soul does indeed require enlightenment in great measure, but it is doubtful whether it is so empty—it could be full of ideas after all! It is this view of Arnobius that Condillac takes up in concurring that the soul needs to be filled from without.

This primordial image or myth that the soul is empty and must therefore be filled from without is still widely, and alarmingly, prevalent nowadays. That is why people are still convinced of their own complete harmlessness, although this couldn't be further from the truth. This idea of being harmless has to do with the idea of the alleged emptiness of the

---

[121] According to Steiner, there are actually seven major embodiments or planetary conditions of the earth system: Saturn, Sun, Moon, Earth, Jupiter, Venus, and Vulcan. During the so-called Sun Evolution, beings called the "Spirits of Wisdom" or "Kyriotetes" began to pour into humanity's body etheric substance, and thus humanity received life, i.e., the "etheric body." Further, the planetary substance condensed to a gaseous state, which emanated light into the cosmos.

[122] From Morgenstern's *Galgenlieder* [Gallow's songs] (1905): *Ein Knie geht einsam durch die Welt. / Es ist ein Knie sonst nichts! / Es ist kein Baum! Es ist kein Zelt! / Es ist ein Knie, sonst nichts. // Im Kriege ward einmal ein Mann / erschossen um und um. / Das Knie allein blieb unverletzt—/ als wärs ein Heiligtum. // Seitdem gehts einsam durch die Welt. / Es ist ein Knie, sonst nichts. / Es ist kein Baum, es ist kein Zelt. / Es ist ein Knie, sonst nichts.* [On earth there roams a lonely knee. It's just a knee, that's all. It's not a tent, it's not a tree, it's just a knee, that's all. In battle, long ago, a man was riddled through and through. The knee alone escaped unhurt as if it were taboo. Since then there roams a lonely knee, it's just a knee, that's all. It's not a tent, it's not a tree, it's just a knee, that's all.]—Christian Morgenstern (1871–1914), famous German poet, writer, and translator (among others, of Strindberg, Hamsun, and Ibsen), best known for his deceptively humorous poetry; incidentally, a friend and adherent of Rudolf Steiner.

[123] Arnobius from Africa (Africanus), around 300 CE. In his psychology and theory of knowledge, he maintained that all knowledge is based upon experience and sensation. Initially the soul is empty, only the idea of God is inborn. The soul itself is of a bodily nature, and achieves immortality only through the grace of God. God Himself, however, is immaterial and eternal. Cf. Francke, 1878; Röhricht, 1893.

soul: If anything evil can be found in it, it must have entered it from the outside! Therefore, someone else must be immediately held responsible for this, be it their father, mother, or schoolteacher. But the soul is not a *tabula rasa*, it is already filled with good and evil when we come into the world. As a matter of fact, there are all kinds of things in it, though we may remain unconscious of this. How else can we account for the fact that the child's mind is full of mythological ideas?

The notion that the soul enters man exclusively through baptism is the Christian concept upon which the rite of baptism is founded. Anatole France's book, *L'Île des Pingouins* [*Penguin Island*], rests upon the same faith. In it, the old abbot St. Maël, half-blinded by the reflection from the polar ice, baptized a school of penguins, thereby causing an enormous dogmatic problem in heavenly circles as to whether it was not a blasphemous act, for only human beings have immortal souls. A council was held in Heaven, but feeling ran high, and no decision was reached. Eventually, Saint Catherine was called upon, and woman's wisdom solved the question with a Solomonic judgement: both parties would be right. Penguins, being birds, cannot have immortal souls; yet it was also true that through baptism immortality is attained. Therefore, she asked the Good Lord to give them a soul, but merely a small one—*Donnez-leur une âme, mais une petite!*[124]

La Mettrie and his *Man a Machine*, and the works of Condillac, gave rise to a reaction against materialist psychology. Among the first to take issue with materialism were, in Switzerland, Jean-Jacques Rousseau (1712–1778),[125] and also Charles Bonnet (1720–1782) in Geneva.[126] In his principal work, *Essai analytique sur les facultés de l'âme* [Analytical essay on the faculties of the soul] (1760), Bonnet took a strange point

[124] In the original: *[J]e vous supplie, Seigneur, . . . de leur accorder une âme immortelle, mais une petite* [Lord, I entreat you to grant them an immortal soul—but a little one] (France, 1908, p. 39). This satirical book by Nobel Prize–winning French author Anatole France (1844–1924) describes the further history of this colony of great auks, mirroring the history of France and Western Europe. Jung also quotes this passage in his commentary to the *Tibetan Book of the Dead* (1935b, § 835) and in *Mysterium Coniunctionis* (1955/1956, § 227).

[125] Jean-Jacques Rousseau (1712–1778), the philosopher, writer, and composer from Geneva. His views were condensed, not quite exactly, in the clichés of "Back to Nature!" and the "Noble Savage." His pedagogical, as well as his political thoughts had a wide impact. See Rousseau, 1755 [1754], 1761, 1762a, 1762b, and his autobiography (1782–1789). There are various references to Rousseau in Jung's works, e.g., on the conflict between the individual and the social function (cf. *Types*, §§ 120–123, 133–134).

[126] It was only in 1815 that the city and canton of Geneva became part of the Swiss Federation.

of view, namely a psychophysical approach to the soul. He held that it was neither purely spiritual nor purely corporeal, but stands oddly in between. To characterize this peculiar intermediate position, he resorted to an image, namely the concept of ether, that is to say, matter that is not matter and yet fills space. Consequently, he conceived of the psychophysical soul as having an ethereal body, in which memories are stored.[127]

Rudolf Steiner, as you know, attributes to all people an ethereal, and also an astral body.[128] Steiner discovered this concept in Indian philosophy, but not so Bonnet. Indian philosophy was unknown at the time, for it is was only later, in the early nineteenth century, that Anquetil Duperron brought the first translations of the Upanishads to Europe, thus opening a new world to the West.[129] Bonnet's concept of the ethereal body can be traced to medieval philosophy, specifically to its primitive idea of the *subtle body*,[130] something similar to smoke or air, the vital spirit[131] that dwells in us. Within the material body, it claims, there exists another, re-

---

[127] Charles Bonnet (1720–1793), Swiss natural scientist and philosopher; an empiricist, early adherent of evolution theory, and discoverer of parthogenesis in aphids or tree lice. In the *Essai analytique*, he developed his views regarding the physiological conditions of mental activity, writing, for example: *Nous ne conoissons pas de Matiere plus mobile, plus subtile, que celle du Feu, ou de l'Ether des Philosophes modernes. C'est donc une Conjecture qui n'est pas dépurvue de probabilité, que l'Organe immédiat des Opérations de nôtre Ame, est un Composé de Matiere analogue à celle du Feu ou de l'Ether* [We do not know of a more mobile, more subtle matter than that of fire, or of the ether of the modern philosophers. Thus, it is a not improbable conjecture that the immediate organ of the operations of our soul is a compound of matter, analogous to that of fire or of ether] (1760, p. 236).

[128] Steiner spoke of three "births" of humans that succeeded one another in cycles of seven years. Up to the age of seven the child is woven within the ethereal and astral cover. After the child's second dentition, the ethereal body is born; at the age of fourteen the astral body, or the body of sensation, is revealed; and at the age of twenty-one the "body of I" is set into spiritual life.

[129] Abraham Hyacinthe Anquetil-Duperron (1731–1805), French traveler, orientalist, and translator; pioneer of *Avesta* studies, but above all, translator (into Latin, from Persian) of a collection of the *Upanishads*, together with lengthy commentaries (1801, 1802), eliciting widespread interest. Arthur Schopenhauer, who played an important role in the kindling of this interest, called it "the production of the highest human wisdom" and "the most satisfying and elevating reading . . . which is possible in the world: it has been the solace of my life, and it will be the solace of my dying" [*die belohnendeste und erhebendeste Lektüre, die . . . auf der Welt möglich ist: sie ist der Trost meines Lebens gewesen und wird der meines Sterbens sein*] (1851, vol. 2, § 184).

[130] This expression in English in the lecture notes: "[T]he philosophers . . . took their *prima materia* to be a part of the original chaos pregnant with spirit. By 'spirit' they understood a semimaterial pneuma, a sort of 'subtle body'" (Jung, 1939 [1937], § 160, pp. 98–99). Jung detailed his views on the concept of the subtle body one and a half years later in the Zarathustra seminar (1988, pp. 441–446, 449–450). Cf. Mead, 1919.

[131] *Lebenshauch*, literally breath of life. The usual expression in Latin texts of the era is *spiritus vitalis*, which was used as a model for this translation.

sembling breath, filling the former, and bringing it to life. This is said to be the breath of life of the heavenly messenger. In the last breath of the dying person, this breath leaves the body. Hence the Indian custom that the eldest son must bend over his dying father to inhale this last breath.

Probably, this idea also informs cannibalism. Ladies and gentlemen, cannibalism, however, is not practiced just for the fun of it! Nor is it due to a lack of meat, nor the cultivation of cuisine. In actual fact it is magic. Through its practice, one gains *prāna*,[132] that is to say, the vital energy of the dying person. I thus confer upon myself the intelligence of the enemy, by spooning out the brain, or his courage, by eating the heart. There was not the slightest doubt among American Indians that one had to swallow the heart of a courageous enemy while it was still twitching.

These ideas had a profound influence also on philosophy. Thus Bonnet, as I have pointed out, made use of the concept of the "subtle body," and also claimed that memory images are stored in the ethereal body. There is an Indian concept that is quite similar, namely an ether known as *akasha*.[133] The term designates a general world ether in which the experiences of all humanity are stored and that are now so material that theosophy claims that they can perhaps even be photographed. Indian philosophy speaks of the "Akashic Records."[134]

Such an autonomous revival of the Indian imagery in Bonnet is an example of palingenesis.[135] Something quite similar can be found in the work of the French philosopher Henri Bergson: the idea of the *durée creatrice*.[136] This idea apparently derives from the material of philosophical deliberation and experience. In reality, however, it constitutes the revival of an ancient idea common among Neoplatonic thinkers. One of them, Proclus, thus asserted: "Wherever there is creation, there is time."[137] But

---

[132] *Prāṇa*, Sanskrit for "vital life," the notion of a vital, life-sustaining force of living beings and vital energy. Its most subtle material form is the breath.

[133] *Akasha*, Sanskrit for ether (also space or sky); the basis and essence of all things in the material world, the first material element created from the astral world.

[134] The Akashic records, a Hindu concept, are described as containing all knowledge of human experience and the history of the cosmos, as a universal "library" in a non-physical plane of existence. The concept was popularized in the West by theosophy and anthroposophy.

[135] From Greek *palin* = again, and *génesis* = birth, creation.

[136] Jung repeatedly referred to the French philosopher Henri Bergson (1859–1941), and his notions of the *élan vital* and the *durée créatrice* [creative duration], the latter being "really the Neoplatonic idea of Chronos as a god of energy, light, fire, phallic power, and time" (1984, p. 428), thus "an example of the revival of a primordial image," which can be found already "in Proclus and . . . Heraclitus" (1919, § 278).

[137] Proclus (ca. 412–485), head of the Platonic Academy in Athens, leading Neoplatonist. In his system, the first principle is "The One," which produces all Being, but is itself

there is no reason to find Bergson guilty of plagiarism. The idea simply resurfaced in him.

In contrast to palingenesis, the autonomous revival of an idea in another epoch, cryptomnesia[138] is a hidden memory, the revival of something that we once knew, but then completely forgot consciously. An example of this is Benoit's novel, *L'Atlantide*.[139] Its entire concept had appeared long before in a book of an Englishman, in Rider Haggard's novel *She*.[140] Benoit was subsequently accused of plagiarism but pledged that he had never read *She*—which is quite possible in the case of a Frenchman.[141]

I myself have discovered such a passage in Nietzsche. Upon making enquiries with his sister, Mrs. Elisabeth Förster-Nietzsche, I learned that the young Nietzsche had read this book when he was eleven years old and later repeated a passage from it verbatim.[142]

After Bonnet, the development of French philosophy was interrupted by the French Revolution.[143] This momentous event was not a sud-

---

beyond being. The divine intellect, the second principle, is outside of time, and produces the soul, which in turn produces Body, the material world. That which *is*—perpetual being—is "remote from all temporal mutation," but that which is *generated* (corporeal formed nature, body), "is either *always* generated, or at a *certain time*" (Proclus, Book 2, 1234).

[138] A term coined by the Genevan psychologist Théodore Flournoy (1854–1920): "by *cryptomnesia* I understand the fact that certain forgotten memories reappear without being recognized by the subject, who believes to see in them something new" (*Planet Mars*, p. 8). Flournoy's work had a great impact on Jung; see the chapter on him in *Memories* (only in German edition, and translated in the 1994 edition of Flournoy's book), and Jung's detailed discussion of Flournoy's book, *Des Indes à la Planète Mars*, later in these lectures (**pp.** 99 ff.). Both Bergson and Flournoy were friends of William James, another important source for Jung (see ibid.)

[139] Pierre Benoit (1886–1962), French novelist. In *L'Atlandide* (1919), two French officers are captured by the monstrous Queen Antinea. She has a cave wall with 120 niches in it, each of them destined for one of her lovers. When all 120 have been filled, Antinea will sit atop a throne in the center of the cave and rest forever. One of the officers is unable to resist her charms, and kills his comrade at her request. Ultimately, he is able to escape.

[140] Sir Henry Rider Haggard (1856–1925), English writer of adventure novels. His *She* (1886–1887) is one of the best-selling novels of all time. It describes the journey of Horace Holly and his ward Leo Vincey to a lost kingdom in Africa. There, they encounter a mysterious white queen, Ayesha, who reigns as the all-powerful "She," or "She-who-must-be-obeyed."—Jung frequently cited *She* and *L'Atlantide* as classical descriptions of the anima figure (see the General Index to the *CW*).

[141] After having been accused of plagiarism, Benoit sued for libel, claiming, among other things, that *She* had not been translated into French, and that he did not read English—which would rule out cryptomnesia, as also alleged by Jung elsewhere (e.g., 1939a, § 516; 1964 [1961], § 457). He lost the lawsuit just the same (*New York Times*, 24 July 1921).

[142] See Lecture 6 of 24 November 1933, in which Jung treats this incidence in detail.—Friedrich Nietzsche (1844–1900) was enormously important to Jung; see, e.g., *Memories*, pp. 123–124, and the seminar on "Nietzsche's Zarathustra" that he started in the summer semester of the following year (1988 [1934–1939]).

[143] In 1789.

den eruption in the external world, but had long been prepared for by philosophers and psychologists. For the great upheavals make their first appearance in the realm of the mind. The idea comes first, and reality follows. For instance, it took Ludwig Büchner's[144] *Force and Matter* twenty years to achieve its breakthrough. It first appeared in the 1870s, and in the 1890s it became the most widely read book in the German lending libraries for workmen.[145] Thus, twenty years lapsed from the birth of an idea to its widespread impact. This shows that it is by no means immaterial which thoughts a university or a schoolteacher hatches, for history might well seize upon this idea. In the case of the French Revolution, this fact was self-evident.

---

[144] (Friedrich Karl Christian) Ludwig (Louis) Büchner (1824–1899), German philosopher, physiologist, and physician; exponent of nineteenth-century scientific Materialism, and founder of the German Freethinkers League.

[145] In fact, Büchner's book, *Kraft und Stoff*, first appeared in 1855, not in the 1870s. With its unabashedly materialistic views on the indestructibility of matter and force, and the finality of physical force, it immediately raised a storm of criticism in the German press, and caused the loss of his teaching post in Tubingen, but was an instant bestseller, and was translated into many languages (e.g., sixth German edition 1859; English translation 1864; 21 German reprints within 50 years).

Lecture 4

10 November 1933

Submitted Questions

I have received an objection from the Catholic side, namely that man does not obtain his soul when he is baptized but already when he is conceived.

The remarks which I made last time about Christianity and baptism seem to have raised misunderstanding. I did not say that Catholics deny the existence of a soul until the child is baptized. Catholic doctrine holds that the human soul comes into existence when the individual is created, that is, at the moment of conception. Some Church fathers assumed that the soul only entered the human embryo forty to sixty days after conception. Personally, I am convinced that not only people but also animals have souls. But what I wanted to say is that baptism entails quite considerable changes in the human soul. Catholic Church doctrine maintains that man is born into a lamentable condition, namely original sin, which is a negative condition. This view coincides with the doctrine of the predicament of underage and nonbaptized children. Should these children die unbaptized, they are caught in a state of being penalized because of original sin.[146] Such thinking makes plain that Catholic doctrine considers the human soul to be in need of salvation. I am not discussing the question of dogma, which does not belong in these lectures. But it is my deepest conviction that all these doctrines are in some manner extraordinarily true. My personal view is that no religious truth is relative, but that each is true in itself. There is no logical standard of comparison. Experiences exist in their own right. These are true and genuine psychological experiences.[147]

[146] According to the Catholic concept (not doctrine) of the *limbus infantium*, unbaptized children die in original sin. They are not assigned to the Hell of the Damned, however, but to the Limbo of the Infants (*limbus* = edge, boundary, the "edge" of hell). This theological hypothesis was officially renounced by Pope Benedict XVI.

[147] Similarly, Jung stated his belief, in *Psychology and Alchemy*, "that God has expressed himself in many languages and appeared in divers [*sic*] forms and that all these statements are *true*" (1935/1936 [1943], § 18; ital. in orig.). He also stressed his conviction that

For the primitives that I encountered the rising of the sun was a religious experience. Were I to criticize these matters, I would be guilty of incredible stupidity from the outset!

***

We are working our way painstakingly through the prehistory of psychology. In the last lecture, I concluded my discussion of French psychologists at the time of the French Revolution. Revolutions are always symptoms of a great mental transformation. You will see this also happen in today's Germany, and you will see that something new will arise out of this transformation.

While psychologists now became obsolete in France, they continued to exist in Germany. Johann Friedrich Herbart (1776–1841), for instance, followed the work of the English psychologists Hume and Hartley, and, like these, developed a psychology of association. Herbart's principal idea is that single elements fuse into a whole, consciousness, through association, following the principle of the attraction and repulsion of ideas.[148] He is the father of more recent physiological and experimental psychology, which I shall not discuss here.

Herbart is followed by Fechner and Wundt (1832–1920).[149] With the latter, a peak is reached. Gustav Theodor Fechner (1801–1887) is the founder of a new psychological point of view, namely so-called psychophysics, which played an essential role in the development of psychology.[150] Fechner at first touched upon an aspect of psychology

religious matters are "not a question of belief but of experience. Religious experience is absolute; it cannot be disputed" (1939 [1937], § 167). Most famous is probably his answer, in the 1959 "Face to Face" interview, to the question of whether he believed in God: "I don't need to believe. I know" (in *Jung speaking*, p. 428).

[148] Johann Friedrich Herbart (1776–1841), German philosopher, psychologist, and educator; known mainly as the first "dynamic" psychologist and the father of scientific pedagogy. He developed a concept of a mechanics of *Vorstellungen* (ideas or representations), which he conceived as dynamic forces that interact, and come into contact, with each other, through the processes of blending, fusing, fading, and combining in a multitude of approaches. Unconscious ideas, when strong enough, can break through a *limen* to consciousness and enter the mind, the "apperceptive mass," through association with ideas already present. Re. Herbart on psychology, see Herbart 1816, 1824–1825, 1839–1840. Cf. Jung, 1946b, § 350.

[149] On Wundt, see note 78.

[150] Gustav Theodor Fechner (1801–1887), German physicist, experimental psychologist, and natural philosopher, also author of poems and literary pieces (under the pseudonym of Dr. Mises); founder of psychophysics (cf. Fechner, 1860); 1834–1839, professor of physics, from 1843 until his death professor of natural philosophy and anthropology at Leipzig University. In his old age he became an exponent of animatism and panpsychism (see below).

that does not suggest that he would make yet another contribution to psychology. His work, *Elements of Psychophysics* (1860), involved not only logarithms but also differential and integral equations! In any event, it had a thoroughly depressing effect on me as a student.[151] Fechner's study is based on Weber's law,[152] which later become known as the "Fechner-Weber law." The law states that the relative noticeable difference between two stimuli corresponds to the relative difference in sensation intensity.[153] It suggests that the ephemeral nature of the soul can be penetrated, and understood, with the help of physical measuring instruments. Although Fechner compiled tables and made calculations, his law is valid only within certain limits. For when the ratings either rise or decrease slightly, it no longer holds true.

If Fechner had solely written *Elements of Psychophysics*, then he could be safely ignored, and his works left firmly shelved. But he was also a so-called philosopher and wrote a series of books besides the *Elements*, whose titles already show that they are of a quite different nature: already in 1836 there appeared *Das Büchlein vom Leben nach dem Tode* [*The Little Book of Life after Death*], then, in 1848, *Nanna oder das Seelenleben der Pflanzen* [*Nanna, or about the Soul-Life of Plants*], and in 1851 *Zend-Avesta oder Über die Dinge des Himmels und des Jenseits* [*Zend-Avesta, or Concerning Matters of Heaven and the Hereafter*].

I presume these titles conjure up the mental somersault of a conflicted mind prone to moving sharply from one extreme to another. In these books, Fechner professes his personal metaphysical convictions. They are, however, nothing but psychology. These philosophical works advance a universal psychophysical parallelism: the psyche is nothing other than the interior, that is to say, the "self-appearance," of events and processes. The body is the exterior, or the so-called extraneous appearance, of the psyche.[154]

[151] Jung had called "Fechner's psychophysics . . . an acrobatic attempt to jump over its own head" (1926 [1924], § 162).

[152] The law of Ernst Heinrich Weber (1795–1878)—who had been Fechner's teacher at Leipzig University—states that the just-noticeable difference between two physical stimuli is proportional to the magnitude of the stimuli.

[153] According to this further development of Weber's law in 1860, the magnitude of sensation is, essentially, the *logarithm* of the magnitude of the physical stimulus, i.e., for the intensity of a sensation to increase in arithmetical progression, the stimulus must increase in geometrical progression.

[154] More precisely, for the mind-body problem Fechner advocated an ontological monism, but a dualism of attributes or properties, depending on the perspective of the observer; in other words, mind and body are two different *Erscheinungsweisen* [modes of appearance] of one and the same *Grundwesen* [fundamental essence or being]: "The

Fechner's great achievement is his distinction between an empirical inner world and an empirical outer world. He speculates even further in assuming that not only the human body but also all living bodies, or any body per se, possess an "interior," that is to say, "self-appearance." For instance, Mother Earth is animated and possesses a soul, which is a by far more comprehensive being than the human soul. She conducts herself like the soul of an angel that embraces all human souls. The totality of human brains thus constitutes the brain of the Earth soul. The highest, omniscient essence of the Godhead is the soul of the universe.[155] While this train of thought is not interesting as philosophy, it is as psychology. Fechner confesses his feeling that his individual soul is not isolated, but is part of a greater whole, which would more or less correspond with the whole of the earth. This is the first occurrence of the notion of a superordinate psychic connection, a connection that is not contained within the individual soul and can be deduced solely by the intellect.

Carl Gustav Carus (1789–1869),[156] a physician and philosopher, worked along almost the same lines as Fechner. In some respects, their thinking is remarkably similar, on account of a dark law according to which things have a tendency to occur at one and the same time, completely independently of causality and space. Thus, simultaneity between stylistic eras occurs all over the world, in Europe and in China, and so

material, bodily, corporal, and . . . the psychical, the mental, are two modes of appearance of one and the same essence; the former being the extraneous mode of appearance for others"—the *Fremderscheinung*—"the latter the inner one for oneself"—the *Selbsterscheinung* or self-appearance; "the two are different because one and the same thing appears in different form when it is perceived by different persons from different viewpoints" [*Das Materielle, Körperliche, Leibliche und . . . [das] Psychische, Geistige sind zwei Erscheinungsweisen desselben Wesens, ersteres die äußere für andre Wesen, letztere die innere Erscheinungweise des eignen Wesens, beide deshalb verschieden, weil überhaupt ein und dasselbe verschieden erscheint, je nachdem es von verschiedenen aus verschiedenem Standpunkt aufgefaßt wird*] (Fechner, 1879, chapter XXII). Cf. Fechner, 1860, vol. 1, pp. 4–6; Heidelberger, 1993, 2010.

[155] Fechner held that mind suffuses the universe, and postulated a "world-soul" or "world-mind" of which everything is a part. He maintained "that the whole universe is spiritual in character, the phenomenal world of physics being merely the external manifestation of this spiritual reality. . . . Atoms are only the simplest elements in a spiritual hierarchy leading up to God. Each level of this hierarchy includes all those levels beneath it, so that God contains the totality of spirits. Consciousness is an essential feature of all that exists. . . . Fechner regarded the earth, 'our mother,' as such an organic besouled whole" (Zweig, 1967). Cf. Fechner, 1861.

[156] Carl Gustav Carus (1789–1869), German polymath, physician, painter, natural philosopher, and philosophical pioneer of the psychology of the unconscious; also distinguished for his work on comparative anatomy. Carus was an important figure to Jung, who claimed: "My conceptions are much more like Carus than Freud" (in *Jung Speaking*, p. 207), and who mentioned Carus repeatedly in his work (e.g., Jung, 1940, § 259). Cf. Hillman, in Carus, 1846, ed. 1970; Shamdasani, 2003, pp. 164–167.

forth. Carus differs from Fechner, however, in that he never endeavored to be an exact psychologist. He was not an empiricist, but a philosopher and a pantheist, influenced by Schelling. His main achievement was the development of a comparative psychology. In 1846, his book, *Psyche, Zur Entwicklungsgeschichte der Seele* [Psyche, on the developmental history of the soul] appeared, and in 1866, *Vergleichende Psychologie* [Comparative psychology]. His works were reprinted by Diederichs in 1926, with Ludwig Klages serving as editor.[157]

Carus was the first to speak of the "unconscious,"[158] and his writings comprise highly modern points of view on it. For instance, he observed that the "key to the knowledge of the nature of the conscious life of the soul lies in the region of the unconscious."[159] For Carus, the soul is the formative principle of the body; the psyche forms its body. To illustrate the relationship between the conscious and the unconscious, he uses the concrete image of life as an unceasingly winding great stream that is illuminated by the sun, that is to say, by consciousness, solely in one single spot.[160] As the stream bears away many valuable things that remain undiscovered, so many treasures are hidden from us, and the actual dynamics proceed in darkness, that is to say, in the unconscious. Consciousness is a beam of light that falls upon certain spots. This notion echoes Kant, with the exception that the dynamics are missing from the latter's thinking. The key to real psychology can be

---

[157] Diederichs was a prestigious publishing house in Jena, Germany, counting among its authors seven winners of the Nobel Prize in Literature, e.g., Bergson, Hesse, or Spitteler. It exists to the present day, since 2008 under the ownership of Random House. Carus's *Psyche* did indeed appear with Diederichs, but such an edition of *Comparative Psychology* could not be verified. Friedrich Konrad Eduard Wilhelm Ludwig Klages (1872–1956) was an influential German philosopher and psychologist, pioneer of graphology.

[158] As a matter of fact, the term "unconsciousness" [*Unbewußtseyn*] was employed already before Carus, for example, in Ernst Platner's *Philosophical Aphorisms* (1776, §§ 11–19, 25, pp. 5–9). Following Leibniz and Wolff, Platner differentiated between ideas with consciousness as perception and without consciousness as dark and obscure representations, e.g., in the sleeping state. The soul, engaged in an uninterrupted process of bringing forth impressions or ideas, vacillates between consciousness and unconsciousness. Schelling, too, as quoted by Jung in an earlier lecture, spoke of the "eternally Unconscious" in 1800 (see pp. 14–15 and notes 93 and 94). Cf. Grau, 1922, p. 63; Lütkehaus, 2005, p. 2005; Nicholls & Liebscher, 2010, p. 9. With thanks to Martin Liebscher.

[159] *Der Schlüssel zur Erkenntnis vom Wesen des bewussten Seelenlebens liegt in der Region des Unbewusstseins.* This is the opening, programmatic phrase, in spaced letters, in Carus, 1846, p. 1. He also held that "illness can have its actual root *only* in the unconscious soul life" (ibid., p. 432).

[160] *. . . dass man . . . das Leben der Seele vergleichen dürfe mit einem unablässig fortkreisenden großen Strome, welcher nur an einer einzigen kleinen Stelle vom Sonnenlicht—d. i. eben vom Bewußtsein—erleuchtet ist* (Carus, 1846, p. 2).

found only in the dark. Mental illnesses and creativity also originate in the unconscious. Carus regards the unconscious as human will and intelligence assuming a cosmic extent. It is a cosmic will, a cosmic intelligence, which creates things and produces consciousness through the individual's unconscious. This philosophy was later taken up by Eduard von Hartmann.[161]

The next link in the chain is Arthur Schopenhauer (1788–1860), who is a great phenomenon, and whose message for the world is of utmost significance.[162] His main work, *The World as Will and Representation*, was published in 1819. Before Schopenhauer, it was generally assumed that the soul consisted predominantly of conscious processes, and that it could be understood rationally. Out of an unconscious idea the world was made into a very beautiful place, and the individual was considered to be truly and properly organized along rational lines. Evidently, one vainly sought for knowledge about the nature of the world as we experience it in these philosophies. But the genius of Schopenhauer brought the world an answer, for which thousands had groped in vain in the dark, and which remains unaddressed in all these empirical philosophies: the voice of suffering.[163] He is the first to declare that the human psyche means suffering, and not only order and purpose. Contrary to all rational consciousness,

---

[161] See below, and note 171.

[162] Arthur Schopenhauer (1788–1860), the great German philosopher. In his philosophy, the world as it appears to us in the dimensions of causality, time, and space, is only our "representation" (or idea), but its essence is ultimately acausal, aspatial, and atemporal "will"—which he equates with Kant's thing-in-itself—manifesting itself in the will to live and to reproduce. The intellect is the will's servant, created to help the will fulfill itself. Only under exceptional conditions can the intellect free itself, and hold up a mirror to the world and to the will itself, or even negate the will, leading to asceticism (not Stoicism) and holiness. Schopenhauer repeatedly named Kant and the Upanishads (which he read in Duperron's Latin translation from the Persian) as crucial influences. Jung had started reading Schopenhauer, and then Kant, as an adolescent: "This philosophical development extended from my seventeenth year until well into the period of my medical studies" (*Memories*, p. 89). He referred to Schopenhauer repeatedly in his talks before the fraternity of the Zofingia (Jung, 1983), and his work had a critical impact on him: "Schopenhauer was so to speak the first man I had encountered who spoke *my* language" (*Protocols*, p. 303). "He was the first to speak of the suffering of the world" (*Memories*, pp. 87–88). Cf. also Jung, 2012 [1925], p. 4. On Jung and Schopenhauer see, among others, Falzeder, 2015; Jarrett, 1981; Liebscher, 2014; Nagy, 1991; Shamdasani, 2003, pp. 173–174, 197–199.

[163] *The World as Will and Representation* abounds with statements such as: "existence itself is . . . a constant sorrow," "constant suffering is essential to life," "every biography is the history of suffering," "the world exists . . . in infinite suffering," "a world whose whole existence we have found to be suffering," "incurable suffering and endless misery [are] essential to the manifestation of will, the world," etc. etc. (7th ed., pp. 345, 365, 418, 426, 528, 531) In short, pain and suffering are always "positive," whereas joy and pleasure are always short-lived, and can be defined only "negatively," by the formers' absence.

he pointed to a gaping fissure that traverses the human soul—namely, between the intellect on the one hand, and a blind will to existence and creation devoid of intelligence on the other. He might as well have referred to this "will" as the "unconscious." For him, however, this creating will is always chaotic and demonic, whereas Carus's notion of the creating will borders on beauty, sweetness, and tedium. Schopenhauer conceives this as the tragic conflict between our consciousness and a dark, profound, and persistent will full of suffering. He thus raises an issue in the psychological discourse that we should never lose sight of, especially because it concerns modern man to a particularly great extent.

As in Carus, Schopenhauer's later treatises became more didactic, such as *Transscendente Spekulation über die anscheinende Absichtlichkeit im Schicksal des Einzelnen* [Transcendental speculation on the apparent deliberateness in the fate of the individual][164] and *Über den Willen in der Natur* [On the will in nature].[165] On balance, however, he continued to conceive the world as an accidental creation, in which the intellect preserves order. The essential purpose of the intellect, he observes, is to hold up a mirror to this blind, demonic creative will, which perpetually gives birth to new worlds, so that it can understand its own nature, and negate its own existence.[166] This peculiar, pessimistic philosophy is strongly influenced by Eastern thought, particularly through Duperron, who brought Eastern teachings to Europe.[167]

---

[164] A chapter in Schopenhauer, 1851. Just as events and figures in our dreams—whether pleasant or painful—are unconsciously composed by ourselves, our lives, says Schopenhauer, seem like the features of one great dream of a dreamer, in which all the other dream characters dream, too, so that everything is linked to everything else, and every seemingly coincidental, unimportant, or untoward event in our lives, if seen with hindsight, appears to have been part of a consistent order and plan. There seems to exist an "inexplicable unity of the coincidental and the necessary, which turns out to be the secret controller of all things human" (p. 200; my trans.). Jung later wrote that it was this treatise "which originally stood godfather to the views I am now developing," that is, his views on synchronicity (1952, § 828).

[165] In this book (1836, 2nd ed. 1854), Schopenhauer used the latest findings of the natural sciences in support of his theory of the will. Schopenhauer himself found this work particularly important and had a special fondness for it.

[166] There are several passages in which Schopenhauer states that the intellect, although originally only a servant of the will, can free itself from this bondage, and become, especially in the case of a genius, "the clear mirror of the world" (1844 [1909], p. 144). In addition, he maintains that it exists also "to hold up the mirror to our will" (1851 [2007], p. 81). Ultimately, the goal of Schopenhauer's philosophy is the *Verneinung* (negation or denial) of the will (cf. 1844, chapter 48).

[167] See note 129 and below.

We see a similar dethronement of Christian views for the first time in the French Revolution, by the enthronement of the *Déesse Raison* [Goddess of Reason] in Notre Dame,[168] in place of the Christian God, from which the entire *Encyclopedia*[169] and the Enlightenment ensued. Evidently, this was not simply the conceit of some few whimsical minds, not a singular eruption of fashionable psychology, but enjoyed the support of all enlightened, progressive minds across Europe. Thus, a shot was fired that resounded in the whole world, and completely shattered medieval man's certainty of the world. Never before had Christianity been publicly denounced, and this blow shook the foundations of the Church. Forces were unleashed that could no longer be captured in the old forms and brought under the yoke.

In this hour of upheaval and destruction, however, the human instinct achieved a compensatory feat: a Frenchman, Anquetil Duperron (1731–1805), went to the East in order to seek the truth there. It was as if Europe had been a psychological being that looked for a new hope in place of the one it had lost. Duperron became a Buddhist monk[170] and translated the Upanishads into Latin. The first rays of Eastern light poured into the cracks made by the French Revolution, and, as France had destroyed, so it was France who first brought something new and living to broken hopes. Schopenhauer let himself be influenced by this message, and translated it into a language, into a philosophy, which the West could understand.

Schopenhauer was followed by Eduard von Hartmann (1842–1906).[171] While he was strongly influenced by Schopenhauer, Hegel, and Schelling, his philosophy derives directly from Carus. Von Hartmann considers the unconscious to be the unity of the will and of the idea as well as the

[168] During the French Revolution, on 10 November 1793, the Goddess of Reason was proclaimed by the French Convention, and celebrated in Notre Dame de Paris. The Christian altar was dismantled, and an altar to Liberty was installed.

[169] The *Encyclopédie*, compiled and written by a group of eminent eighteenth-century writers—the *encyclopédistes*, among them Rousseau and Voltaire—edited by Denis Diderot and Jean le Rond d'Alembert (whose brilliant "Introduction" became famous in its own right; 1751). They stood for the advancement of science and secular thought, tolerance, rationality, and the open-mindedness of the Enlightenment.

[170] This is unlikely; there is no mention of this in the standard biographies of Duperron.

[171] (Karl Robert) Eduard von Hartmann (1842–1906), German philosopher, best known for his book on *The Philosophy of the Unconscious* (1869), a bestseller at the time (10th ed., 1890). He argued that his concept of the unconscious reconciled the extremes of the logical idea (Hegel) and the blind will (Schopenhauer), the unconscious being both will and reason (idea), and the absolute all-embracing ground of all existence.

purposefully acting foundation of the world of divine and absolute nature. Psychology is scant in his writings; he was rather a philosopher and wrote about the *Philosophie des Unbewussten* [*The Philosophy of the Unconscious*] (1869).[172]

In the meantime, a new development was underway in France. Maine de Biran (1766–1824)[173] also acknowledges the existence of an unconscious sphere, whose main features largely coincide with those of consciousness, however.

Two other thinkers follow in his footsteps: Ribot[174] and Binet.[175] The latter's point of view about the unity of the soul is interesting. His *Alterations de la Personnalité* [Alterations of the personality] are modern in a certain sense, since he proceeds not from details but from the human personality as a whole.[176]

---

[172] There are numerous references to von Hartmann in Jung's works. In 1952, Jung stated that it was Kant, Schopenhauer, Carus, and von Hartmann who "had provided him with the tools of thought" (in *Jung Speaking*, p. 207).

[173] Marie François Pierre Gonthier (Maine) de Biran (1766–1824), French philosopher, member of the Council of Five Hundred (1797), and councilor of state (1816). Originally informed by sensualism, he then rejected Condillac's and others' view of knowledge as derived solely from sensation, replacing it by a combination of voluntarism ("*Volo ergo sum*") and spiritualism. His work was important for Bergson, who considered him the greatest French metaphysician since Descartes and Malebranche (Bergson, 1915, p. 247). Most of his writings were published only posthumously in various editions (cf. Maine de Biran 1834–1841, 1859).

[174] Théodule Armand Ribot (1839–1916), French psychologist and philosopher. Professor of comparative and experimental psychology at the Collège de France (1888); founder of the *Revue philosophique* (1876), and organizer of the first international congress on psychology (1889). He initiated the study of a positivistic and physiologically oriented psychology in France. Ribot's work was significant in that it represented the beginnings of pathological psychology, including neuropsychology. Among his students were Pierre Janet, who succeeded him at the Collège de France, and Alfred Binet; but his work was also important for thinkers such as Henri Bergson.

[175] Alfred Binet (1857–1911), French psychologist. Widely known for inventing the first usable intelligence test (1905; Binet-Simon test), later refined by Lewis Terman as the Stanford-Binet test (1916), which itself is the basis for a test used still commonly today.

[176] In his book (1892)—which he dedicated to Théodule Ribot—Binet deals with those alterations of the personality that result in "the division or dismemberment of the self," that is, cases in which "the normal unity of consciousness is broken up and several distinct consciousnesses are formed, each of which may have its own system of perceptions, its own memory, and even its own moral character" (p. x). For Binet, "'personality' is a thing of relative synthesis which may be manifested in very varied degrees of completeness" (J. Mark Baldwin, in ibid., p. vi). Jung quoted many examples of Binet's book in his doctoral dissertation and repeatedly referred to him in his works, although he did not mention Binet's distinction between the types of "introspection" and "externospection" as one comparable to his own typology (cf. Brachfeld, 1954; Ellenberger, 1970, p. 703; Jung & Schmid, 2012, p. 13).

In France, Binet is followed by Pierre Janet[177] und Liébeault,[178] and in the United States indirectly by William James (1842–1910).[179] The latter's *Principles of Psychology* (1890a) place him at the forefront of psychology, and contributed to advancing it by steering it away from academic circles towards personality research and towards medical doctors.

We thus enter the field that my introduction has been leading up to. History, as you know, has always chronicled individual lives and psychologies, particularly of outstanding persons and "great men"; and among these, the "men of action" have predominantly attracted the interest of psychological historians.[180] But there exist also other personalities besides such "men of action"—"psychic" people, people marked by their inner experience. They stand out much less, and yet we possess authentic historical sources about them, and find them in a place where we would have hardly thought to encounter them: in the lives of the saints, the

[177] Pierre Janet (1859–1947), French psychologist, physician, philosopher, psychotherapist, and hobby botanist. Director of the Psychological Laboratory of the Salpêtrière under Charcot (1889), and successor of Ribot at the Collège de France (1902). Many of his ideas—e.g., on the "subconscious," fixed ideas, automatisms (Janet, 1889), dissociation, obsessions, inferior and superior parts of the human mind, his method, which he called "psychological analysis," or his trauma theory of psychic disorders, which "has given rise to a whole theory of neurosis and psychosis by the subconscious persistence of an emotional traumatism" (Janet, 1930, p. 128)—resemble the (later) ones by Freud. In the ensuing priority dispute, Janet lost out at the time, but "[s]ubsequent historical research has tended to support [his] claims" (Shamdasani, 2012, p. 37; cf. Ellenberger, 1970). Jung studied with Janet in Paris in 1902, and drew much from his work: "I did not start from Freud, but from Eugen Bleuler and Pierre Janet, who were my immediate teachers" (Jung, 1934 [1968], § 1034). Jung's work abounds with references to Janet's notions, such as *fonction du réel*, *abaissement du niveau mental*, *dissociation*, *sentiment d'incomplétude*, *formes inférieures et supérieures* (of mental life), or *idées fixes subconscientes*, and he repeatedly acknowledged his debt to him.

[178] Ambroise-Auguste Liébeault (1823–1904), French physician and pioneering hypnotherapist, intellectual father of the so-called Nancy or Suggestion School, of which Bernheim became the leading proponent (in contrast to the rival school of Charcot in Paris). Cf. Ellenberger, 1970.

[179] William James (1842–1910), the famous Harvard psychologist and philosopher, elder brother of the writer Henry James. Jung had first met James at Clark University in 1909, and paid him a visit the following year. In an unpublished chapter of *Memories*, he states that they had an "excellent rapport," "that James was one of the most outstanding persons that he had ever met" and that he became "a model" for him (in Shamdasani, 2003, p. 58). "There are only a few heaven-inspired minds who understand me. In America it was William James" (*Jung Speaking*, p. 221). On James and Jung, see also Taylor, 1980.

[180] Such as Karl Gotthard Lamprecht (1856–1915), for whom history was "nothing but applied psychology" (1905, p. 29); or Wilhelm Dilthey (1833–1911), who first differentiated between *Geistes-* and *Naturwissenschaften*, a distinction still widely used in German-speaking countries, for which there is no exact equivalent in English (roughly: humanities or social "sciences" and natural sciences): While natural sciences "explain" nature (*erklären*), social sciences "understand" humans (*verstehen*).

Acta Sanctorum,[181] in the court records of witch trials, and later in the miraculous accounts of stigmatized individuals[182] and of somnambulistic persons. By the late eighteenth and the early nineteenth century, a fairly copious literature had emerged on these strange personalities.

One such account is Justinus Kerner's (1786–1862) *Die Seherin von Prevorst* (1829) [The Seeress of Prevorst].[183] This is not a work of literature, but actually a case history, that is to say, an account of a curious and remarkable "psychic" personality. Kerner was a practicing physician in Württemberg. He wrote the history of his patient Friederike Hauffe in 1829. Nobody seems to have thought of bringing this story in a line with modern psychology, but we will see that it contains some highly interesting psychic phenomena. The subtitle of the book is *Eröffnungen über das innere Leben des Menschen und über das Hereinragen einer Geisterwelt in die unsere* [Revelations concerning the inner life of man, and the inter-diffusion of a spirit-world in ours], which shows what stirred Kerner's interest in this case, namely an objective, substantial world of spirits. Kerner's psychology thus still very much has the character of the projected, for it is not the spirit-world that interfuses with our world, but the unconscious. Kerner attended to this "Seeress" for many years, and offered an affectionate account of all the strange events surrounding her.

---

[181] *Acta Sanctorum* [Acts of the Saints], 68 folio volumes of documents examining the lives of Christian saints, organized according to each saint's feast day, a mammoth undertaking from 1643 up to 1940, probably making it the editorial project of the longest duration.

[182] That is, those who had received the stigmata, like St. Francis.

[183] It should be noted that the English translation of *Seeress* by Catherine Crowe (1845) is a quite "free translation, giving the sum and substance of the book" (in ibid., p. ix), but leaving out long passages. Justinus Andreas Christian Kerner (1786–1862), German physician and poet, noted for his literary work, for his discovery of botulinum—the "fatty toxin" in sour sausages (now used in botox)—and above all for his work on spiritism, occultism, and somnambulism, as in the book quoted by Jung. It is the record of his treatment of Friederike Hauffe (1801–1829), a young, depressed, and severely ill woman. In her magnetic states, she saw ghosts, the souls of other people, etc., and described the spirit world and how it interacted with "our" world, all of which Kerner recorded with great care. The book—"the first work in psychiatry devoted to a single patient" (Shamdasani, 2012, p. 32)—caused a sensation and also great controversy in Germany, and became a bestseller. Jung read it in 1897 (ibid., p. 31), and it had a significant effect on his own experiments with his cousin Helene Preiswerk that formed the basis of his doctoral thesis (Jung, 1902). After he had given Helene a copy of the book, she became convinced that she herself was a reincarnation of Hauffe.

Lecture 5

17 November 1933

Justinus Kerner's *The Seeress of Prevorst* is not a case history in a modern sense, but as it were a dubious account of one of the peculiar and romantic lives that were quite common at the time. Kerner belonged to the school of Romanticists. He was not a scientist,[184] and his book contains a series of more or less naive observations and interpretations. So please do not think that I subscribe without reservation to anything and everything that my deceased colleague Dr. Kerner tells in his book!

Kerner describes the case of his patient Friederike Hauffe, who was born in Prevorst near Löwenstein in the German state of Württemberg in 1801. Her father was a forester. She had a number of siblings who suffered from infantile spasms, eclamptic fits,[185] which occur frequently during teething. In those days, these spasms could already be distinguished from epileptic fits, so we may assume it was not epilepsy. More important is another detail about her family history, namely, that her grandfather possessed what is called "second sight" in Scotland, that is to say, a strange perception of things that did not exist at the time, but which then, curiously enough, actually came true.[186]

The Seeress was a normal, healthy, and happy child. Soon, however, it was noticed that she had a great number of colorful and graphic dreams. What struck people was that these dreams often came true. For instance,

[184] This may be too harsh a statement, however, in view of Kerner's medico-scientific researches (e.g., the discovery of botulinum).

[185] MS: *Gichtern* = an obsolete term for eclamptic fits; mistranslated in Kerner, 1845, p. 32, as "gout" (= *Gicht*). This condition was frequent among children at the time, also due to unhealthy and unhygienic food, and a not uncommon cause of death. In a village near Prevorst, Neuhütte, for instance, that "sort of St. Vitus's dance" became even "epidemic, chiefly amongst young people" (*Seeress*, p. 31).

[186] Kerner mentions a few instances of paranormal experiences of Johann Schmidgall, Friederike's grandfather, into whose care she was put as a child (e.g., seeing specters or apparitions; cf. ibid., pp. 16–19, 34–35), but none of second sight. On "second-sight," see ibid., pp. 85–88.

her father once reproached her for mislaying an object. The child felt innocent, and beheld in a dream the place where her father had left the item.[187] Now this is doubtless a most naive story, and the case might strike us as suspicious. We might well be tempted to suspect that the child had hidden the object herself before she ventured to find it again. Without doubt, such strange things, like a great many others, do happen in our world! On the other hand, we might also be doing the child an injustice by mistrusting her account. In such cases, we are well advised to bide our time and wait to see whether such events repeat themselves. Other strange events occurred, however. The child began to play with hazel rods and soon proved to be a good diviner. Divination was popular among farmers at the time, and she had probably observed some of them looking for the location of water veins.

Her grandfather would often take her out for walks. On these occasions, he observed that she would begin to freeze terribly, tremble all over, and become frightened in certain places. In some cases, he was able to prove that she reacted thus when they crossed ancient burial sites. Later, the child could no longer endure church, "because there were ancient tombs beneath the church floor," and she had to go to the galleries because she could not endure sitting in the choir directly above the graves. Matters, however, became even more serious. She developed a sense of uncanny places, and she would see figures in places that were said to be "haunted."

> Thus, there was an apartment in the Castle of Löwenstein—an old kitchen—which she could never look into or enter without being much disturbed. In the very same place, some years afterwards, the spectre of a woman was, to her great horror, seen by a lady, who had never been informed of the sensations experienced by the child.

In itself, this report proves nothing, of course, for it could have been simply a fear of ghosts that brought about her visions. Her thoughts corporealized, however, even though she had not thought them. And this is a fact: no thought can take a bodily form once it has been thought, because by that point it has already been thought away, so to speak. If I imagine myself thinking in a dark room that it could be haunted, then it is definitely not haunted, because I have already thought away that thought. But if we do not entertain the thought, then it could well occur. There are also cases of a simultaneous "double vision." Kerner continues:

[187] This, and the following examples, are taken from ibid., pp. 33–35.

> To the great regret of her family, this sensibility to spiritual influences, imperceptible to others, soon became too evident; and the first appearance of a spectre to the young girl was in her grandfather's house. There, in a passage, at midnight, she beheld a tall, dark form, which, passing her with a sigh, stood still at the end of the vestibule, turning towards her features that, in her riper years, she well remembered. This first apparition, as was generally the case with those she saw in after life, occasioned her no apprehension. She calmly looked at it, and then, going to her grandfather, told him that "there was a very strange man in the passage, and that he should go and see him"; but the old man, alarmed at the circumstance—for he also had seen a similar apparition in the same place, though he had never mentioned it—did all he could to persuade her that she was mistaken, and, from that time, never allowed her to leave the room at night.

It could be argued that her grandfather's influence induced the girl to have such strange visions. But it is probably more accurate simply to assume that she, too, possessed this gift of "second sight," that is to say, that she could actually see thoughts. The grandfather tried to reason her out of her belief in what she had seen, but he was unable to shake her conviction of the reality of these experiences. Kerner did not doubt that she really did see ghosts because he himself was convinced of their existence.

If you want to relate to such people, you will, of course, have to take it for granted that there are such things. Telling people who believe in ghosts that "There are no ghosts!" is futile. We have to meet them at their own level; if we do not, or if we immediately question or even mock what they believe, we will throw away any advantage. In any case, we can make no sweeping assertions in this field, for all proof is lacking—we do not know whether or not ghosts exist. It happened to me that I attended the palaver of some highly respectable Negroes. Naively I asked them whether they had ever seen a ghost. They all averted their gaze and looked as if I myself had just conjured up the most frightful specter.[188] One should not mention ghosts, for they are the unspeakables. Even uttering the word is

[188] On Mount Elgon, "in East Africa, . . . during a palaver, I incautiously uttered the word *selelteni*, which means 'ghosts.' Suddenly a deathly silence fell on the assembly. The men glanced away, looked in all directions, and some of them made off. My Somali headman and the chief confabulated together, and then the headman whispered in my ear: 'What did you say that for? Now you'll have to break up the palaver.' This taught me that one must never mention ghosts on any account" (Jung, 1950, § 759).

perilous. This holds true even today—there are certain matters that you must treat only with kid gloves.

From all this, we may conclude that the girl inherited her grandfather's disposition, and that she also possessed the gift of exteriorization, that is to say, she was able to "externalize" psychic processes. Such processes are based on psychological facts. From the point of view of science, the question whether or not ghosts exist is far from answered. Quite possibly, Kant's prophecy, uttered in his *Dreams of a Spirit-Seer* (1766),[189] will still become true:

> [I]t will be proved . . . that the human soul also in this life forms an indissoluble communion with all immaterial natures of the spirit-world, that, alternately, it acts upon and receives impressions from that world of which nevertheless it is not conscious while it is still man and as long as everything is in proper condition.[190]

This statement is both remarkable and most optimistic, since I cannot imagine how the existence of these things could ever be proven. For we are unable to discern whether we are observing processes of the unconscious or something outside ourselves, unless we had exact physical methods to be able to prove objectively reality in this field.

In my opinion, "second sight" is not an illness, but a gift that is not as such pathological—otherwise every other gift would be pathological, too, and we would be obliged to speak of an "intelligence disease," an "art disease," and so forth. All gifts also involve pain, however, not merely pleasure. With regard to her dreams, by the way, the "Seeress of Prevorst" is not extraordinary. Countless individuals have prophetic, anticipatory dreams. There is nothing peculiar about such dreams; quite often they are surprisingly banal.

The first truly pathological symptom afflicting the Seeress was a strange irritability of the eyes.[191] This condition lasted perhaps for a year. Nothing was outwardly wrong with the eyes, so it was probably a psychically induced sensitivity to light. Such a symbolic sensitivity to light is quite common: those people cannot bear the light, they squint psychically, so to speak, and are unable to tolerate a too clear consciousness.

---

[189] Early on in his studies at the university, Jung not only read Kerner's book, but "virtually the whole of the [spiritualistic] literature available to me at the time," of which "Kant's *Dreams of a Spirit Seer* came just at the right moment" (*Memories*, p. 120).

[190] Kant, 1766, ed. 1900, p. 61. This passage was also quoted by Kerner, *Seeress*, p. 82.

[191] A "remarkable sensibility in the nerves of the eye, (without the least inflammation,)" when she was still a "young girl" (ibid., p. 35).

Something makes them shy away from clarity, for instance, because of an unconsciously bad conscience, out of a fear of being found out, either by someone else or indeed by themselves.

Her adolescence was more or less typical. As far as we know, nothing else noteworthy happened thereafter, not until she became engaged to a Mr. Hauffe, a merchant tradesman, who, however, plays only a very shadowy role in the subsequent course of events. On the day of their engagement, the seminary priest T. in Oberstenfeld, whom she greatly revered, passed away; he was over sixty years old and died of natural causes. The Seeress, who was about twenty years old at the time, attended the funeral and could not tear herself away from his grave after the burial.[192] She was evidently in a peculiar state. In actual fact, she had a vision, in which she beheld the deceased hovering above the grave as a ghost—an event that she later reiterated in a poem.[193] Contrary to earlier visions, this one had a tremendous impact on her, and she remained for a long time under its spell.

She married in 1821, at the age of twenty. At first, her life took a normal course; she went through an uncomplicated pregnancy and gave birth to a child. Thereafter, however, in February 1822, she had a dream that proved to be fateful for her. In the dream, she was lying in her bed, with the dead priest lying beside her. Next door she overheard her father and two doctors discussing the nature of what they considered to be a serious illness. She cried out: "Leave me alone by this dead man!—he will cure me!—no physician can!" She felt as if they wished to draw her away from the body, and cried aloud in her dream: "How well I am near this corpse; now, I shall quite recover." Her husband heard her talk in her sleep and woke her up. The following day, she came down with a violent fever that lasted a fortnight.[194] No one knew what it was. The fever led to a severe neurosis, from which she died on 5 August 1829, at the age of twenty-eight.

What happened here? Imagine a patient came to my practice and told me this dream. I would obviously wonder why I should calmly let her rest side by side with the corpse, and ask her: "Why did you come to me at all, since you believe that no doctor can cure you?" If she replied: "The

[192] Ibid., pp. 36–37.

[193] This poem was omitted from the English translation (1845), and is also missing in the most recent German re-edition (2012). In it, she described how she was absorbed, at the gravesite, by the "angel-image" [*Engelsbild*] of the dead priest on the burial mound (*Seeress*, orig. ed., p. 31).

[194] *Seeress*, pp. 39–40.

dream seems strange to me, I cannot imagine why I should think that the dead could cure me," I would take her on as my patient; but if she gave the same answer as in the dream, it would be fatal, and I could do nothing for her. As a matter of fact, such a person would probably never come to analysis, and if she did, she would certainly manage to maneuver the doctor on to the side of death, unless he had great experience in such cases. One might say that the very fact of her coming to analysis would in itself be a considerable argument against her being wholly on the side of her dream. But it is a very ominous dream, and as a doctor I consider it very questionable whether anything could have been done for her. There are cases where it is better not to interfere. We must fulfill our duty as doctors, but the fact remains that some people are not meant to be cured, they are not fitted for life, and if you step in and interfere, fate always takes its revenge on you.

There is no doubt that the Seeress de facto took sides with the dream, and that it thus assumed a fateful meaning. We think we could not possibly allow ourselves to become entangled in a certain fate through a dream, but since her psychology was different from ours she did become ensnared in it. She identified with the dead man and already died as it were while she was still alive. She retreated more and more into this "back-world," until she vanished from the "fore-world." The particular fate of the Seeress became apparent from then on. The death of the old priest was the experience that made clear to her that she would live more with the dead than with the living. What really mattered to her were the encounters with her inner figures, beside which husband and child were mere shadows. She felt healed and normal when she accepted the dream, that is to say, when she slipped back into the psychic background, and she felt ill if she ventured into the real world where she encountered insurmountable conflicts; so she stepped ever further back into the unconscious, until she ceased to exist.

I have chosen this particular case, and am treating it in detail, in order to show you the immense reality of the inner world. There are a considerable number of people whose psychology is somewhat similar, in that from the outset the outer world means less to them than this "back-world." We would assume that such people will ultimately see reason, or we would be inclined to some drastic, no-nonsense treatment. But that would be completely useless. For them, this background is infinitely more real than the outer world. The whole outer world means nothing to them. I have known cases where people became as it were somnambulists and disappeared into the unconscious. It was as if they had never been born completely and could consequently not trust this bright, sun-lit world sufficiently to be able to live in it.

This is not empty madness. Such matters exist in our lives, and a whole life can rest upon such realities. Although we do not notice them as long as all is well, they exist nevertheless. Our consciousness perceives the outer world; it is an organ of perception. But behind our consciousness there stands a perceiving subject, and this is no *tabula rasa*. This subject is not simply another exterior, but instead it comes endowed with a background, with whose help it is able to interpret perceptions in the first place. Human children are not born with empty brain capsules, but instead with a complete brain, created for eons. Consequently, every child is born with a predetermined assumption of the world, of which it is not conscious, but which is nevertheless at work. Failing this innate opinion, we would be unable to grasp the world at all. There is no escape from this psychic background with which we enter life, it can only be accepted. Endowed with it, however, we must comprehend the world according to this disposition.

Here, we must take into account certain tribal influences. Possibly, certain human tribes split themselves off from the common tree at an early stage, and consequently their genetic constitution differs noticeably from ours. Besides far-reaching differences, however, some highly remarkable parallels exist, too.

The fact that a thought can assume shape is a primitive fact: the primitive is incapable of abstract thought. When he thinks, something makes itself apparent within him, usually in the stomach. I experienced this myself once when a Pueblo Indian said to me: "Americans are mad! They say that we think with our heads. But only madmen think with their minds; reasonable people think with their stomachs!"[195] When "it" thinks in him, the Negro's stomach rumbles. Occasionally, something might "sit heavily on our stomach," too, but that would only affect those of us who think emotionally. Since not all thoughts are of an emotional nature, but touch upon our thoughts, they appear, in the case of primitives, in the world around them in projected form.[196]

---

[195] Jung later wrote, however, that Ochwiay Biano ("Mountain Lake"), chief of the Taos Pueblos, had told him that they thought with their *hearts*, not stomachs (*Memories*, p. 276).

[196] This paragraph refers either to remarks made at the end of this, or at the beginning of the following lecture, conceivably in answer to a question from the audience. There is no reference to this in Jung's own notes, nor in those taken by M.-J. Schmid.

Lecture 6

24 NOVEMBER 1933

SUBMITTED QUESTIONS

I have received a letter from a lady who writes:[197]

> When you discussed the dream of the Seeress of Prevorst, you said: "As soon as we take sides with a dream, it assumes a fateful meaning." You added that you could not treat a patient who believed in his [*sic*] dream. Now I ask myself: Would it not be somehow possible that this patient could be induced *by the treatment itself* to renounce his belief in the dream? Would not the fact alone that he consults the doctor be an indication that he is actually ready to change his attitude, and that he harbors a faint hope somebody could help him to abandon this belief?

I am utterly convinced that the Seeress would consult a doctor at best because of some secondary symptoms. Her fate illustrates that she no longer desired to live. On the contrary, she converted the doctor to turn towards the darkness with her. I have known such cases myself. Fate proves to be stronger.

The second questions regards cryptomnesia, that is, the revival of memories that we do not recognize as such, and the example for this in Nietzsche.[198] [Jung reads the relevant passage from Nietzsche's *Zarathustra* as quoted in his dissertation:][199]

---

[197] Jung probably did not read the full text of the letter, which is reproduced here in my English translation of the original held in the ETH Archives.

[198] See p. 26 and note 138.

[199] Jung had already used this as an example of cryptomnesia in his dissertation (1902, §§ 140–142), and also in his article on cryptomnesia (1905a, §§ 180–183). He further quoted it in his seminar on Nietzsche's *Zarathustra* (1988, vol. 2, pp. 1215–1218), the seminar on *Children's Dreams* (1987 [2008], p. 17), and in his contribution to *Man and His Symbols* (1964 [1961], §§ 455–456). The passage referred to is from Nietzsche, 1883–1885

> Now about the time that Zarathustra sojourned on the Happy Isles, it happened that a ship anchored at the isle on which standeth the smoking mountain, and *the crew went ashore to shoot rabbits*. About the noontide hour, however, *when the captain and his men were together again, they saw suddenly a man coming towards them through the air*, and a voice said distinctly: "*It is time! It is the highest time!*" *But when the figure was nearest to them* (it flew past quickly, however, like a shadow, in the direction of the volcano), then did they recognise with the greatest surprise that it was Zarathustra; for they had all seem him before except the captain himself . . . "*Behold!*" *said the old helmsman,* "*there goeth Zarathustra to hell!*"

Now, when Zarathustra goes to hell, it is completely irrelevant that the crew has gone ashore to shoot rabbits. Rabbits, of all things! Then I remembered, from my student days, a green book with a red edge from the Basel University library. So I looked for Justinus Kerner's four-volume *Blätter aus Prevorst*, and found the pertaining story in it: "An extract of awe-inspiring import from the log of the ship *Sphinx*, in the year 1686, in the Mediterranean" [*Ein Schrecken erweckender Auszug aus dem Journal des Schiffes Sphinx vom Jahre 1686 im mittelländischen Meere*]:[200]

> The four captains and a *merchant, Mr. Bell, went ashore* on the island of Mt. Stromboli *to shoot rabbits*. At three o'clock *they called the crew together* to go aboard when they saw, to their inexpressible astonishment, two men *flying rapidly over them through the air*. One was dressed in black, the other in grey. *They approached them very closely, in the greatest haste*; to their greatest dismay they descended amid the burning flames *into the crater of the terrible volcano, Mt. Stromboli*.

The two men were acquaintances from London, and when Bell and the naval officers returned home, they heard that both men had died at about that time. I subsequently wrote to Nietzsche's sister. In her reply, she mentioned that when Nietzsche was eleven years old, they visited their grandfather, Pastor Oehler, in Pobles,[201] whose old library held a complete set of Justinus Kerner's writings. They furtively entered the library and illic-

---

[1911], p. 88. The emphasis was added, presumably by Jung himself (or the note-taker), to emphasize the concordances of the two texts.

[200] Kerner, 1831–1839, vol. 4, p. 57.

[201] MS: "Pobler," because Jung misread the name of the village in Elisabeth Förster-Nietzsche's answer to his query.

itly read all the frightful stories contained in the *Blätter aus Prevorst*.[202] I read the *Blätter* myself in 1898; the Nietzsche edition of 1901 had yet to appear, and, that aside, it was not advisable to read Nietzsche in Basel in those days. How exactly this cryptomnesia occurred escapes me. Whenever any classical idea occurs to us, however, it will enter into association with all sorts of mnemonic material, either consciously or half-consciously.

***

In the last lecture, we observed that the "Seeress of Prevorst" identified with the dead priest; [in the dream] she lay down beside him on the bed, as if they were husband and wife, or for that matter as if she were already buried next to him. This suggests that she is more inclined to adjust not to outer events and happenings, but to what arises from within, that is, from the subject. It is dead certain that ghosts and similar phenomena are things that we experience within, as if they did not appear within the field of vision in front of us, but from behind, as it were, from the unconscious. The only guarantee we have that such things do exist is the [evidence of the] I. People are confronted with them through the I, as if something existed behind the I whose source we are completely ignorant of.

Clairvoyance more or less bears out this point: as if those people were able to see around the corner; or they hear things other people don't. Something somehow reveals itself from within—or from "behind"; it does not come from the frontside, not from the clear world of consciousness, and it is not perceived by the sensory organs. This, however, occurs solely in a particular state. The Seeress, too, notices these things only when she is in an exceptional state.

Thus, there are contents that come to us from the outside, and those that come from the inside, as the following diagram illustrates.

---

[202] Jung gave different versions of Elisabeth Förster-Nietzsche's information. While in 1902 and 1905, he correctly quoted her that Nietzsche had been between twelve and fifteen years old when he had read Kerner (1902, § 141; 1905a, § 182), in later accounts he repeated the version as given above, namely, that this happened sometime between his tenth and eleventh year (1964 [1961], § 456; 1987 [2008], p. 17; 1988 [1934–1939], p. 1218). He also somewhat embellished the story by adding that the children's reading was done illicitly. This is not mentioned by Elisabeth Förster, and Nietzsche himself wrote: "My favourite place was granddad's study, and it was my greatest pleasure to browse in the old books and journals" (in Janz, 1978, p. 62; my trans.)—which does not indicate any forbidden activity. On Jung's correspondence with Nietzsche's sister concerning this question see Bishop, 1997.

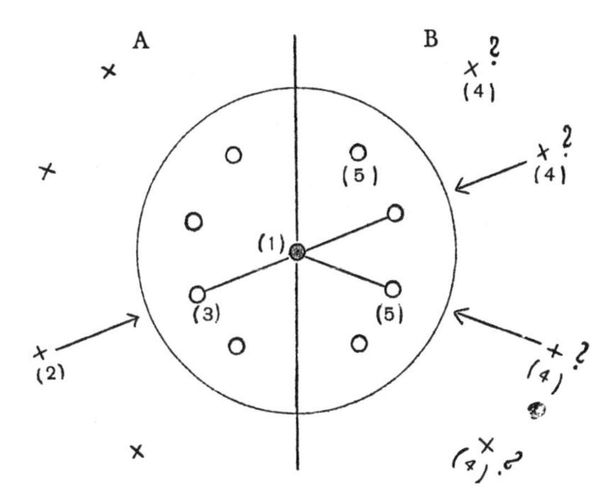

Our consciousness in the center is like a spider in its web, with threads, through which the psychic contents are associated with the I. This inner realm of consciousness is surrounded by outer objects that create an image in consciousness through the sense organs and the nerve tracts. We are dealing here with the objective facts of everyday life, with persons or things in the external world that make an impression on us. A Mr. X., for instance, makes an impact on consciousness and creates a psychic content.

Other impressions reach us from the invisible inner world. We can hypothesize that this background also contains objects, comparable to those in the external world, and that these objects equally leave an imprint in our consciousness, through images that are registered by our psyche. These psychic images are not necessarily conscious, just as many images constellated by outer objects do not reach our consciousness either—they are psychic contents, pure and simple. They are conscious only if they are associated with the I, otherwise they do not exist in consciousness. They can nevertheless act upon me, without my knowing, in this case, what is acting upon me. Thus, the I is in the center and acts like a magnet that attracts all contents.

This "I"—what a peculiar matter it is! First of all, it is something subjective. Nothing seems to be behind it, and yet you are able to think about the "I." We can objectify it and make it the subject matter of our thinking. In speaking to you, I am aware of the fact that I am speaking to you. I can then tell myself: "You did that well," or the opposite. It is as if someone else were standing behind me who observes me; a second I that comments upon the first I's actions. Or it is as if you were in a labyrinth, with mirrors both behind and before you, so that your image is infinitely mirrored, and becomes ever smaller. For instance: I No. 1 is lecturing. I No. 2 hears that I No. 1 is lecturing. I No. 3 hears that I No. 2 hears that I No. 1 is lecturing. This hypothesis can be carried out ad infinitum, although we better not do this. The fact of this second I making its comments, however, is familiar to us all. It is what the English call "self-consciousness,"[203] that is, being conscious of oneself to a high degree. There is someone sitting behind you who can't keep his trap shut, talking about you all the time. When this self-consciousness grows too strong, as in the case of stage fright, for instance, you become inhibited, and a pathological state can arise so that you have to undergo treatment.

Now there is indeed something behind it, and I have referred to this darkness as the subjective factor or the background, the subjective I. In the case of Mrs. Hauffe, for instance, we realize that there are matters "back there" that appear from within and are perceived by the "inner eye." It is not a capability unique to the Seeress, however, to perceive these contents; we, too, are capable of it. Not during daytime, of course, but when you go to bed the most incredible ideas can occur to you. For instance, the idea of burglars enters your mind, and you hear them outside the window, or believe that you see them in the room. These thoughts come from within and are "real." Not with all the will in the world can you get rid of them, although you know exactly that they are imagined: "Doctor, I believe that I have cancer. I obviously know that this is nonsense, but I cannot help myself. The more I resist it, the stronger the idea."[204]

We merely see the contents—that is, strange "realities," obsessive thoughts for example, such as an imagined illness or the like; all we see are the contents, but we have no idea from where they come. It is possible that there are objects "back there" that enter consciousness, but this is merely a hypothesis that cannot be proven, because only the I can bear witness to it. While I can demonstrate that the psychic images in the foreground correspond to a real content, I cannot do likewise for those "at the back."

---

[203] This word in English in the German transcripts.

[204] Jung refers here to the case of "a professor of philosophy and psychology who consulted [him] about his cancer phobia. He suffered from a compulsive conviction that he had a malignant tumor, although nothing of the sort was ever found in dozens of X-ray pictures" (1964 [1961], § 467). Jung mentioned the case repeatedly, e.g., also in 1938 [1954], § 190, in his Terry Lectures (1939 [1937], §§ 12–13, 19–21, 26–27, 35–36), and in his seminar on *Dream Interpretation Ancient & Modern* (2014, p. 71).

It can happen to me that all of a sudden the idea of a burning house occurs to me, and nothing will make the thought go away. And word could reach me at that very moment that it is *my* house that is ablaze. So I think: Aha! Now I have the proof, the verification, of the fact that I have "second sight"—even though it was not my own house that I saw, but merely *a* house; however, it corresponds to my house that is ablaze. I have perceived real events in an unreal manner. And yet I am the sole proof, for no one else has seen the event.

There are a great number of cases, however, in which this "vision" corresponds to nothing in reality. In these cases a subjective factor is at play, a dark point that lies behind us. For instance, a pleasant, reasonable fellow awakes one morning, and suddenly a bad mood overcomes him. Naturally, he is oblivious of the subjective factor, but blames his wife or a badly cooked breakfast—and before long a row erupts between the couple. "Well, how did you know that your wife would annoy you?" "Well, I already knew that beforehand!" Something walked in and seized him from behind, and he projected it from the subjective background onto his wife. People always see these things in the outside world, even when they occur within themselves, and we project every day with the most unbelievable shamelessness. This mechanism is particularly apparent in newspapers, where the thoughts and moods of writers and politicians are projected into others *de l'autre côté de la rivière*;[205] that is where we like to see the devil.

After this somewhat ungainly attempt to elucidate matters at the limits of human cognition, let me return to the "Seeress of Prevorst." In accord with her dream, the Seeress evidently expects to recover her health from what lies entirely in the background: she dreamt that she would regain her health if she lay down beside the dead priest. Thus, she expects her health to lie in this back-world, in the dark sphere. As I have already mentioned, this dream is very ominous; it is so ominous in fact that as a medical doctor I would doubt whether anything could be done. There are indeed cases where one is well-advised not to help. It is not worth it. Obviously, we must do our duty as doctors. But if you manage to heal someone at the expense of your own health, you have fallen victim to a misunderstanding. For the fact remains that there are people who are not supposed to become healthy, since they are not destined for life.

As you will recall, on the day after the dream, the Seeress was stricken by a fever that lasted a fortnight. No one knew what it was. The fever

[205] French, from the other side of the river.

was followed by a neurotic state, what the French call a *grande hystérie*, which resulted in her death, seven years later, in 1829. On 13 February 1822, she had the dream; on 27 February 1822, what were believed to be "severe spasms in the breast" caused her to awaken from her sleep.[206] Probably, this was a nervous heart affection, in effect a "heart cramp." Such cramps can have organic or psychic causes; the latter was certainly true in her case. This is a metaphor, that is, an instance of mimic representation. The body illustrates something that cannot reach consciousness. "My heart cramped up," we say, or "my heart stood still"; it was "as if an iron fist had squashed my heart." Such "heart metaphors" abound in the novels of the writer Adalbert Stifter.[207] The Seeress has received an ominous impression from "behind." Naturally, she projects this condition outward: The nurse had left the water standing recklessly on the washstand, instead of placing it on her bedside table. In reality, it was rather the unconscious knowledge that her death was imminent that cramped her heart.

[206] *Seeress*, p. 40.

[207] Adalbert Stifter (1805–1868), Austrian writer, painter, and pedagogue.

Lecture 7

1 December 1933

In today's lecture, I shall discuss Mrs. Hauffe's symptoms. Her case belongs to the field of mediumistic phenomena. Strictly speaking, these lie outside the field of medicine, and are part of parapsychology. I must mention these phenomena, however, since they do exist and are therefore psychologically important. We can well maintain a critical attitude toward these matters, but we have to heed the facts, keep our minds open, and prevent theoretical biases from obstructing our thinking.

Mrs. Hauffe's condition was first and foremost of a somnambulistic nature: everyday consciousness slipped away from her. Somnambulism[208] is an exceptional psychic state, not a twilight state. For the Seeress, it constitutes a heightened state of consciousness that, for her, is actually the more normal state than the waking state. Such conditions involve great exertion, however, and therefore cannot be sustained for a long time. If she had managed to maintain these states, she would have been a "higher" being.

For instance, she speaks in verse. Other phenomena include visions and hallucinations; for example, she sees a mass of fire in her body.[209] This phenomenon can also be observed, by the way, in the case of ordinary neuroses; the visions are of a symbolic nature. Another phenomenon in her case is autoscopy:[210] for instance, she was lying in bed while seeing

[208] Here, not sleepwalking, but an altered state of consciousness, a kind of "sleep-*waking*" state, or "magnetic sleep," a condition, however, we "must not call . . . sleep—it is rather a state of the most perfect vigilance" (*Seeress*, p. 24). Already Armand de Puységur (1751–1825), a disciple of Mesmer (see below) and a pioneer of magnetism, described how people in that state were able to diagnose their illness, to foresee its course, and to prescribe the correct treatment.

[209] "At one time, she spoke for three days only in verse; and at another, she saw for the same period nothing but a ball of fire, that ran through her whole body as if on thin bright threads" (ibid., p. 44).

[210] The experience of seeing one's own body from a vantage point or position outside of the physical body.

herself sitting beside it,[211] that is, she beheld an exteriorized image of herself. This is not unusual in such cases, and also occurs with the seriously ill and dying.

After that, her eye affliction recurred, and again she shunned light.[212] Outer light was painful to her, and so she concentrated on the inner light. She no longer looked out of the front door, so to speak, but out of the backdoor, into the depths of the subjective world, thus bringing about other positive manifestations of the unconscious, of the background. She saw all kinds of things which she projected into the outer world as ghost figures—some related to her, some related to other people. The ghosts provided her with treatment, notably with Mesmeric sayings[213] and magnetic manipulations. She frequently had a double vision of other people, since behind their personality, perceptible through the senses, there stood another that bore the qualities of the soul.[214]

Her condition worsened rapidly on account of all these apparitions. She subsisted on a very poor diet; and when she came to Kerner in 1826 she was already in a very bad state, suffering from malnutrition and scurvy.[215] She had no appetite, which evinces a deficient will to live and corresponds to a sentiment of death, as we can also see in cases of melancholia. Fasting is also a technique in asceticism and yoga; deadening the drives depotentiates the outer world so that the vision can be directed purely inward.

As I mentioned, the Seeress came to Kerner in 1826. He attended to her and examined her as best he could, that is, in a most naive and primitive way. For instance, he firmly believed the nurse's report of being unable to bathe the patient, since she would float on the water and could not be submerged:

---

[211] Ibid., p. 44; cf. pp. 57–58.

[212] Ibid., p. 44.

[213] Franz Anton Mesmer (1734–1815), German physician, and famous, charismatic, and highly controversial healer. Noted for his theory and practice of "animal magnetism." Kerner was well acquainted with his ideas—he himself had been cured from a nervous affliction, as a boy, by a "magnetic" cure—and in his old age paid homage to Mesmer in a monograph (Kerner, 1856).

[214] "When Mrs. H– looked into the right eye of a person, she saw . . . the picture of that person's inner-self. . . . If she looked into the left eye, she saw immediately whatever internal disease existed . . . and prescribed for it" (*Seeress*, pp. 73–74). In other instances, she saw "another person behind the one she was looking at," for example, "behind her youngest sister she saw her deceased brother" (ibid., p. 46).

[215] Her gums had become scorbutic, and she had "lost all her teeth" (ibid., p. 52). The ultimate cause of scurvy—a deficiency of vitamin C—was not known until 1932, and treatment was inconsistent.

> When she was placed in a bath in this [magnetic] state, extraordinary phenomena were exhibited—namely, that her limbs, breast, and the lower part of her person, possessed by a strange elasticity, involuntarily emerged from the water. Her attendants used every effort to submerge her body, but she could not be kept down; and had she at these times been thrown into a river, she would no more have sunk than a cork.

"This circumstance," Kerner continues, "reminds us of the test applied to witches, who were often, doubtless, persons under magnetic conditions; and thus, contrary to ordinary law, floated on water."[216] In Basel, for example, witches were thrown bound by their hands and legs from the Rhine bridge into the waters. Those who had not drowned upon reaching the suburb of St. Johann were certain to be regarded as witches! It makes no difference whether or not this holds true for the Seeress. The fact remains that she affected others in a way so that they believed such things to be true about her.

She also developed a strange sense of, or a particular sensitivity to, the properties of materials, particularly of stones and minerals, which exerted a certain influence on her, thus prompting Kerner to conduct numerous experiments.[217] What this faculty is we do not know. With regard to her visual faculty, too, Kerner observed strange, crystallomantic phenomena in her. Some people, when placed before a mirror or crystal ball, see things that are not there, that is, past, present, and future events; some are actually hypnotized by this. Crystal balls were one of the props employed by medieval sorcerers. They also served divinatory purposes[218] in China. What such people see, if anything at all, are of course processes from their own unconscious.

The Seeress also possessed this faculty when Kerner placed her under hypnosis. He describes one of his experiments thus:

> A child happening to blow soap-bubbles: She exclaimed, "Ah, my God! I behold in the bubbles every thing I can think of, although it be distant—not in little, but as large as life—but it frightens me." I then made a soap-bubble, and bade her look for her child that was far away. She said she saw him in bed, and it gave her much

[216] Ibid., pp. 65–66.

[217] Cf. *Seeress*, chapter 8. Only fragments of this chapter appeared in the English edition of 1845.

[218] That is, the art of prophecy.

> pleasure. At another time she saw my wife, who was in another house, and described precisely the situation she was in at the moment—a point I took care immediately to ascertain.[219]

On another occasion, she looked into a glass of water and saw in it a carriage that did in fact pass them by twenty minutes later.[220] Here I would like to remind you of the *scène de la carafe* in Dumas's *Joseph Balsamo*.[221] These are ancient magical practices, and it is not impossible, after all, that they are true. As I mentioned, however, we must leave aside this question at this juncture.

Not only could the Seeress see with her eyes, but also with the pit of her stomach, in that she could recognize objects placed there. Kerner writes:

> I gave Mrs. H– two pieces of paper, carefully folded: on one of which I had secretly written, "There is a God"; on the other, "There is no God." I put them into her left hand, when she was apparently awake, and asked her if she felt any difference between them. After a pause, she returned me the first, and said, "This gives me a sensation, the other feels like a void." I repeated the experiment four times, and always with the same result.[222]

In a further experiment, Kerner wrote on one piece of paper, "There are spectres," and on another, "There are no spectres." "She laid the first on the pit of her stomach, and held the other in her hand, and read them both."[223] The fact that the visual sense is transferred onto another sense is not unique. Rather, it is said to occur in other cases, such as when one places one's hand upon an object in order to see something. In this regard, I refer to the famous American medium Mrs. Piper, who was William James's medium, and who placed letters upon her forehead in order to read them.[224] In those days, it was believed that such experiments only succeed when those who have written the letters in question were still

---

[219] Ibid., pp. 74–75.

[220] "She . . . saw . . . a carriage travelling on the road to B–, which was not visible from where she was. She described the vehicle, the persons that were in it, the horses, &c.; and in half an hour afterwards this equipage arrived at the house" (ibid., pp. 46–47).

[221] For a detailed description of this "scene of the carafe" see below, pp. 107–108.

[222] Ibid., pp. 75–76.

[223] Ibid.

[224] Cf. James, 1886, 1890b, 1909. Leonora Piper (1857–1950) was a famous American medium, and the subject of investigation, apart from William James, of leading psi researchers at the time.

alive. One of Mrs. Piper's female friends bequeathed a letter to her, depositing it at a bank. When the friend died, the letter was fetched from the bank, and placed with the envelope upon the medium's forehead. She focused on it with utmost concentration, much more than customarily, and eventually she said: "I believe it says thus and so"; the contents, however, were in fact utterly different.[225]

Up until that time, the visions and perceptions that had occurred to the Seeress bore reference solely to outer matters, in other words ones that could ultimately be verified. Then, however, she had a rather curious vision that dumbfounded Kerner, and that used to baffle me, too—namely, the vision of a sun-sphere. In her vision, it assumed the shape of a real solar disc, situated in the region of the stomach or of the solar plexus. The disc rotated slowly and scratched her, thus agitating her nerves.[226] Later she had such a distinct vision of this sun-sphere that she could furnish a very interesting drawing of it.

Four points are indicated on the circle,[227] but we have no knowledge what they mean.[228] The [outer] sun circle is divided into twelve parts or segments, corresponding to the twelve months, that is, the zodiacal circle or the sun year. This sun circle is the sixth one; under it lie five others, and above it a seventh, empty circle.[229] We must picture these things as though they were lying in a horizontal position. This is how the vision appeared. For us the concept of an empty circle is an extremely strange idea, for we have lost touch with such matters, because we are accustomed to constantly directing our attention outward and only rarely do we turn to look inward. Any educated Hindu, however, will immediately recognize the meaning. Since this empty circle does not appear in the drawing, the sun circle represents the outermost circle or the circumference.

---

[225] A detailed description of this experiment is in Sage, 1904, chapter 7 (pp. 52–64).

[226] *Seeress*, p. 114; the chapter on "Spheres" was heavily abridged in the English translation of 1845, so that the mentioned details are missing there.

[227] The following descriptions of the seven circles or rings differ considerably in the various lecture notes, and sometimes they even contradict one another. I have attempted to "extract" a description that is as clear as possible and in accordance with Kerner's account (which is itself rather unclear, as Jung duly notes), which facilitates reading, but also means that this is the passage of which we can least be sure that it is a faithful and more or less literal reproduction of what Jung actually said. The numbering of the circles, too, is inconsistent—sometimes they are counted from the center to the periphery, sometimes the other way around, even by Mrs. Hauffe herself. To avoid confusion, this has been standardized to the second method here.

[228] These are the four "cardinal points"; see the beginning of the next lecture.

[229] *Ich fühle unter diesem Ringe noch fünf solcher Ringe und über ihm noch einen leeren* [I feel that there are another five spheres beneath this one, and above it one other, empty sphere], the Seeress is quoted by Kerner (only in German edition of *Seeress*, p. 114).

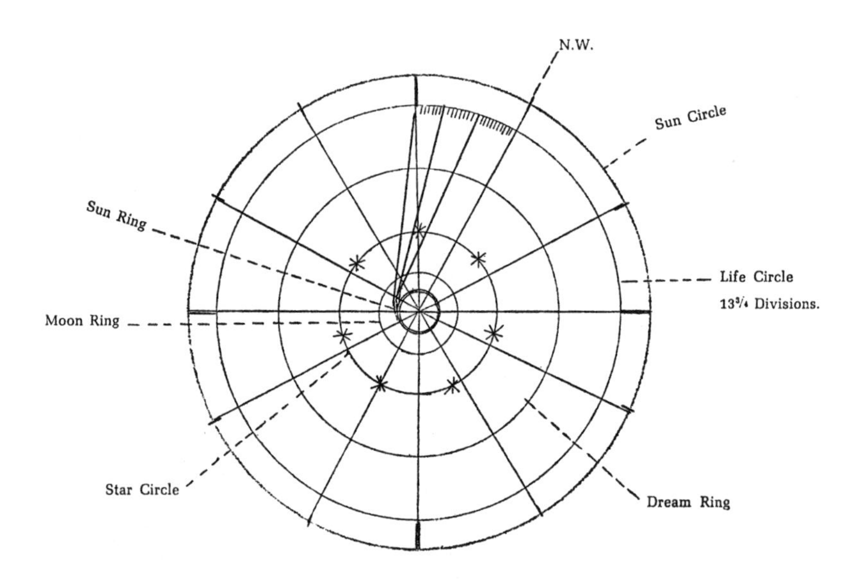

The next [fifth] circle is divided into thirteen and three-quarter segments, which correspond roughly to a lunar cycle of 27.3 days. Thus, the Chinese not only have a solar cycle, but also one that follows the course of the moon. The whole is therefore a kind of time wheel that has sun divisions as well as moon divisions. This fifth or life circle she called her "calendar," upon which she entered minute strokes to record facts or experiences that affected her in a pleasant or unpleasant way: headaches, heart cramp, and so on.

The circles lie under each other and rotate, with rotation commencing in the northwestern point. Thus, this curious calendrical calculation begins in the northwest—and, regarding direction, moves not clockwise but counterclockwise as in the German swastika. Whereas the lines for the month run radially from the sun circle straight to the center, the lines from the life circle are drawn at a tangent, so that what emerges is not a rotating but rather a spiraling movement.

Next comes the fourth circle, which is also divided into twelve segments.[230] She referred to it as the "dream ring" or the "realm of the souls of animals." This is difficult to explain, since Kerner's account is unclear. These three constitute the outer circles.

[230] The description of the spheres and the following quotes are in *Seeress*, ed. 2012, pp. 114 sqq., and ed. 1845, pp. 114–115, 130–131.

The third circle, which is the outermost of the three inner circles, is characterized by a division into seven segments; it is the circle of the seven stars. Next comes the second circle, the "moon ring," and finally the innermost circle, the so-called sun ring, which is bright and shines like a sun. In this center, there is also the spirit and the truth. She called it the "bright mid-point," the "sun of grace." Of this, she caught merely a fleeting glimpse, during which she beheld her "conductress" or protecting spirit [*Führerin*]; she believed, furthermore, that with her many other ghosts had peered inside.

The Seeress said that the sun circle was like a wall, through which nothing could reach her.[231] Outside lay the bright everyday world, where life was dreadful, and she therefore preferred to remain confined within these rings. In her drawing, she depicted people as little hooks, pointed and unpleasant. She felt them as blue flames on the outer ring, not at all corporeal, but only ideational. For her, Justinus Kerner was the mediator between herself and the dangerous outer world.

About the first, central circle she said that she would feel comfortable there, looking from there into the world, the paradisiacal world, that is—by which she means the inner world—in which she had once been. She asserted that the [second] moon ring was a cold and bleak sphere, representing the abode of the souls in the "intermediate region." She shuddered to think of it and was scared of it. She had only vague recollections of it because out of fear she "only swam hither and thither over it." From there, the souls entered either the sun or the stars. This, too, is a strange idea, unless we consider it in the light of history: since time immemorial the moon was thought of as *receptaculum animarum*, the seat of the souls. Manichaeism explains the waxing and waning of the moon each month as its filling with the souls of the deceased until it is filled completely, turns towards the sun, gives the souls to it and thus is on the wane again. Then a new circle begins. This idea was brought from Beijing to southern France through the heretic teachings of the Albigenses.[232] Today, this obviously strikes us as an old tale, but such matters remain pertinent. One fantasy claims that after the [First] World War all souls arrived on the moon and fertilized it so much that grass began to grow there.[233]

---

[231] *durch die nichts an mich konnte* (p. 115). The English translation is misleading: "like a wall, beyond which she could not move" (p. 130).

[232] The Albigenses, or Cathars, represented the most influential Christian heretical movement in the Middle Ages, thriving in some areas of Southern Europe, particularly in northern Italy and southern France, between the twelfth and fourteenth centuries.

[233] In general, "[a]ccording to the ancient belief, the moon is the gathering-place of departed souls" (*Transformations*, § 487), "the abode [*receptaculum*] of departed souls"

We see that, in the case of the Seeress, we are dealing with the revival of an archaic image, a palingenesis, unless, that is, somebody had insinuated that idea to her in her past, although this is quite unlikely in view of her strictly Catholic environment. The notion that souls wander from the moon to the stars is not new, either: stars have been associated with birth and death since time immemorial. Meteors are souls. Or when a Roman Caesar died the astronomers had to find a new star in the sky to account for his soul. We also find this idea in the Indians and in poetic language. The fact that there are seven stars corresponds to a mythological conception: seven is a sacred number, just as all cardinal numbers from one to nine are sacred, in different ways in different peoples.

The reason for this is that primitives can solely count to ten, since they can count only their 2 × 5 fingers. For instance, Swahili[234] has only five numerals; numbers greater than five are designated by "many," whether this be six, a thousand, or a hundred thousand. Before the outbreak of the First World War, rumor had it in East Africa that one thousand[235] German soldiers had marched through the region. Troops were committed to investigate the incident, and it turned out that only a patrol of six men had been seen! Failing a corresponding word for the actual number, the corporal who made the sighting had simply reported "many." Many: *nyingi*; very many: *nyingi sana*. But when very many are meant, perhaps a million, their voices imitate the throwing of a stone: *very* many: *nyingy sa-a-na—a*. Besides, primitives associate numbers with geometrical figures: for example, 2 = II; 3 = III or Δ; 4 = IIII or □. Their conception of numbers is entirely unarithmetic. Thus, for instance, two matchsticks plus one matchstick does not equal three matchsticks, but two two-matchsticks and one one-matchstick. The number is a quality that inheres in the matchstick. A "three-matchstick" cannot turn into a

(Jung, 1927 [1931], § 330). The claim that the moon was so fertilized by the many souls of soldiers killed during the First World War that a green spot appeared on it goes back to the Greek-Armenian mystic, writer, composer, and choreographer, George Ivanovich Gurdjieff (1866? –1949). According to Gurdjieff, the moon feeds on souls, and when the moon is very hungry, there are wars. Jung quoted this story also in his seminar on children's dreams (Jung, 1987 [2008], p. 167). In Barbara Hannah's English edition of these lectures, however, and in contrast to all other lecture notes, this belief of Gurdjieff's is rendered differently: he would have been "convinced that the *spots on the sun* are caused by the unusual number of souls that migrated there during the war, and I [Jung] have met two doctors who firmly believe him" (p. 35; italics added).

[234] Jung had learned some Swahili for his stay at Mount Elgon, and "manage[d] to speak to [the natives] by making ample use of a small dictionary" (*Memories*, p. 294).

[235] In Hannah: "10,000" (p. 35).

"four." Such thinking results from the magical character of all perceptions among these people.

Therefore they can count without being able to count. For instance, they base counting on images, such as the one that grows from a herd or the size of the plot of land occupied by a herd. One native owned a herd of approximately 150 cows, which he would count every evening. But he could count only to ten. He knew the name of each cow, however, and immediately noticed when a particular cow was missing.

Lecture 8

8 December 1933

Submitted Questions

The first correspondent doubts that the example mentioned in the previous lecture—Nietzsche—represents actual cryptomnesia.[236] In response to this question, I would state that I did not use Nietzsche's example as proof.

The second correspondent enquires whether human consciousness is identical to the sum of knowledge about psychic processes? In response, I would maintain that knowledge is self-evidently consciousness. Everything associated with the I is, of course, conscious.[237]

***

In the last lecture, we considered the Seeress's vision of several spheres. Let me give you a summary. The outer circle is the sun-sphere, which she describes thus: "This ring has 12 parts, and in it I see the main impressions of what occurred to me at the time."[238] This circle simultaneously denotes the year and symbolizes the entire world. Now imagine the so-called life-sphere lying closely superimposed thereupon; it is divided into thirteen and three-quarter parts, according to the lunar cycle. From it emanate the radiuses, which are actually not radiuses, because the lines go tangentially to the innermost circle.[239]

The three inner circles are, starting from the center: 1. the solar center—the innermost small circle radiates like a sun; 2. the moon circle;

[236] The question is from Harry R. Goldschmid from Zurich (letter of 29 November 1933; ETH Archives).

[237] Barbara Hannah notes that "There is another question asking if the ego is identical with what we know? To answer this question would lead us too far, it will be answered to some extent as the lectures proceed."

[238] This was not translated in the (incomplete) English edition, which also does not contain many of the following quotes. In these instances the translations are our own, and the page references are to the reprint of the German original (ed. 2012, p. 114).

[239] Ibid., p. 120.

3. the seven-starred star circle. "Seven" is not necessarily an arithmetical term, but rather it represents a quality. Then comes the circle of the souls of animals, or the dream-ring. Evidently, she assumes that a certain identity exists between the nature of the dream and the nature of the animal soul.

Kerner reports the words of the Seeress: "Under this ring I feel five other such rings, and above it an empty one."[240] The circles are thus layered, as it were. Regarding the sun-sphere, she says:

> The real, bright day and the people lie for me outside of the great ring, and I see more or fewer of them in the various sections. I prefer to represent these people as checkmarks. I feel the spirit of all people with whom I associated, but I do not feel or know anything of their body, their name, etc. Likewise (she said to me), I cannot think of you as a man or as a body, of you least of all. I always feel you as a blue flame going around and around the outer ring . . . , together with your wife in the same ring. But she is in human form, and more to the outside. . . . [241]

Characteristically, she has no corporeal conception of man, but an ideational one. She disavows the corporeality of man and, in accordance with her overall attitude, ascribes reality solely to what lies within. In so doing, she depotentiates the outer world, and the inner world thus assumes reality.

> This outer ring with the blue flame circling around it is like a wall to me, through which nothing can reach me. I myself am in the ring. When I imagine myself outside of this ring I feel terrible and get scared. When I imagine myself to be free within the circle, however, I get kind of homesick.[242]

She identifies the outer orbit with Justinus Kerner. His blue flame is moving there. These two act like a wall. Kerner serves her as protection against the outer world, as a mediator as it were. She feels as if she were confined within the ring. It is like a magic circle that has been drawn around her: outside lies the world of anxiety and fear, within the positive life. This circle coincides with the reality peculiar to her that surrounds her. She dissociates her consciousness from the outer world.

---

[240] Ibid., p. 114.
[241] Ibid., pp. 114–115.
[242] Ibid., p. 115.

Thereafter, she speaks about the small orbits at the center, enumerating these from the outer rim inwards:

> I felt well in the first ring at the center (I feel as if seven stars stood above it). From there I talked to the world, in which I had once been. . . . [243]

She does not say: "In which I *am*." Evidently, she considers this to be the world in which she *has been* rather than the one where she is now. Although she lives in this world, to her it appears as some ghostlike world, a kind of recollection or illusion, whereas her inner world represents reality. Leaving aside interpretation for the moment, let us simply listen to her words. To her, there appears to exist a world outside in which she once lived, that is, a prenatal world, a paradisiacal world, a celestial world, from which she was born and was thereafter impelled to leave—much like an expulsion from paradise. So here, too, the story of the Fall of Man, or the fall of the angels, is repeated.

> In the second ring I was cold and shivering; it must be a cold world. I never talked there. I just kind of swam hither and thither over it, and a few times I looked into it. I do no longer know what I saw there, I am afraid whenever I think of it. It is terribly cold and bad there. This ring has the light of the moon.[244]

This second orbit is treated like a vague recollection—she merely "swam hither and thither over it." Similar ideas occur in ancient mythology and in the teachings of the ancients about the becoming of the soul. We will discuss these in greater detail later. Just as the souls of the dead ascend from the earth to the moon, into the *receptaculum animorum*,[245] so they descend again through the moon-ring at birth.

When the Seeress stands in the outermost orbit, she effectively stands in the land of the dead, and looks beyond the severing cold ring of the moon into the sun, that is, the innermost sphere:

> The third ring is as bright as the sun, but its center is even brighter. I saw an impenetrable depth in it, and the deeper it was, the brighter it became. I never went there, I was only allowed to look into it. I would like to call it the sun of grace. . . . There, in the utmost clarity

[243] Ibid., p. 116. The translation gives this in reported speech: "In the outer orbit, over which seven stars seemed to shine, she was at ease and happy; she spoke into the world from it . . ." (pp. 130–131).

[244] Ibid.

[245] Latin, the receptacle of souls.

> of the innermost ring, I saw my guide, and from there I also received the prescriptions, although I no longer know, how.[246]

Here the concept of a female conductress or guide appears for the first time. With somnambulistic persons and psychics, we always encounter such guiding figures. They act as guardian angels, or "protecting spirits," attending to the well-being and woes of others. In the case of female somnambulists, they are very often male figures, and vice versa. There are very famous cases, such as Mrs. Piper, who had not just one such figure but a whole group of "controls," a veritable small-sized general staff, which was in constant attendance.[247]

Evidently, the mid-point is the ultimate depth, that is, the radiant fullness of light. Within its radius stands the guiding principle, which in the case of the Seeress finds expression in the spirit of her grandmother. Psychologically speaking, within our interior there stands a guide. We all carry unconscious guidance within us. Even if we think we have been in control, we experience time and again that sometimes it is not we who decide, but rather something inside us that decides the outcome. All peoples of the earth believe that there exists another being that directs us and determines our conscious decisions.

This being is also the guide of the Seeress. Guidance proceeds from the centerpoint. This point, however, is not situated within the center of consciousness, but within the solar plexus.[248] It has been called thus since antiquity, following the discovery that in a state of ecstasy we are able to see this light through "sympathy," that is, through the sympathetic nervous system. There are peoples—and in Greece there was even a sect—who practice omphaloscopy, that is, the contemplation of the navel, in order to experience this inner vision.[249]

The Seeress furnished a second drawing of the life-sphere, in which it stands alone. As we shall see, this second illustration casts new light

---

[246] Ibid., pp. 116–117.

[247] On Mrs. Piper, see Lecture 7 and note 224. Jung mentioned this group of five of her psychic "controls," called "Imperator," in his commentary on *The Secret of the Golden Flower* (1929, § 60). Cf. Hyslop, 1905, pp. 113 ff.

[248] In anatomy, the *plexus solaris*, or solar plexus, is a complex of radiating nerves situated near the celiac trunk, superior mesenteric artery, and renal arteries. It governs various inner organs as well as the contraction and slackening of smooth muscles. In the Hindu chakra system it is known as the *Manipura chakra* or the ten-petaled lotus.

[249] Omphaloscopy, the contemplation of the navel, was practiced as a form of meditation to experience God as light by the Hesychasts, named after Johannes Hesychastes (454–559). Hesychasm (from the Greek *hesychia*, stillness, rest, quiet, silence) is an eremitic tradition of prayer in the Eastern Orthodox and Catholic Churches.

on the other spheres. It is filled with foreign writing; this is the language of the ghost country. So in this case we encounter the phenomenon of glossolalia,[250] that is, the occurrence of a foreign language entirely unrelated to any known language. Such occurrence is a frequently observed phenomenon with introverted individuals who have retreated far into the background.

The inner life-sphere has certain features, upon which the Seeress comments as follows:

> In the center of this circle there is something that posits numbers and words, and this is the spirit. Just as this world lies within the sun circle, there lies a completely different and higher one in this life sphere (soul), and this why everybody has a premonition of a higher world. . . . From there, the spirit looks into the center of the sun circle.[251]

We must be mindful of the fact that these spheres are lying one upon the other. The life-sphere is thus situated underneath the sun-sphere, and its midpoint is thus no longer identical with that of the sun-sphere. What is called the sun at the top of the sun-sphere is known as the spirit in the life-sphere. We may find this completely crazy at first. I have found out over the years, however, that such questions have preoccupied the human mind for centuries. The Seeress continues: "The center of the life sphere is the seat of the spirit, and this is its right and true place."[252]

This center, to which truth is attributed, exists in other systems, too. One remarkable parallel, albeit not the only one, is the Dharmakaya,[253] the so-called divine body of truth, in Mahayana Buddhism.[254]

Regarding the second sphere, the Seeress gives the following account:

> The second circle represents already the beginning obfuscation of the spirit with regard to what is good, but in a way that it is still able to return to the better of its own accord.

---

[250] Speaking in tongues.

[251] *Seeress*, pp. 120–121.

[252] Ibid., p. 121.

[253] Dharmakaya, also known as the "Truth Body," is a central concept in Mahayana Buddhism, according to which the originally enlightened Buddha transformed the five poisons of the spirit (ignorance, hatred, envy, pride, and greed) into the five underlying aspects of wisdom.

[254] Mahayana Buddhism ("Great Vehicle") is one of two major schools of Buddhism, together with Tibetan Vajrayana Buddhism. This is distinct from Theravada Buddhism on account of its "great" motivation, which strives for liberation not merely for itself, but for all suffering creatures.

> The third circle represents a diminished level of what is good, but still in a transitory stage, so that it is still free to return completely to the inner circle. The third circle is also the last one of the spirit.[255]

There is obviously a gradation, in that this central point, which denotes the spirit, loses gravitas and intensity through emanation. The spirit starts to degenerate in the second sphere, and once it reaches the third sphere, its intensity is even less. The same applies to the sun-sphere. The sun loses intensity through radiation into outer space. It constitutes a sort of emanation system.

In the third sphere, Mrs. Hauffe sees the numbers, upon which her entire calculation was based, the numbers 10 and 17.[256] I mention this merely to illustrate that number mysticism, which is always related to inner systems, begins here. Extended systems also exist that begin with such number mysticism. In other systems, such as the Chinese one, 10 is the number of the earth; 17 remains undocumented (10 + 7), but for the Seeress it is the spiritual number. The life-sphere is a lower ring. Consequently, the spirit is to her a phenomenon that lies beneath the sun-sphere. Thus, it does not represent the highest level.

For the moment, I shall not dwell upon this symbolism any further. Instead, I would rather resume my discussion of the symptomatology of the Seeress by considering other strange phenomena. She frequently experienced instances of clairvoyance, specifically clairvoyant perceptions and clairvoyant dreams. Here are a few examples:

> One night she dreamt that she saw her uncle's eldest daughter go out of the house with a small coffin on her head: seven days afterwards died her own child, aged one year, of whose illness, at the time, we had not the least idea. She had related the dream to me and others on awaking.[257]

Furthermore:

> Another night she dreamt that she was crossing some water, holding in her hand a piece of decaying flesh, and that, meeting Mrs. N–, the latter had anxiously inquired what she was going to do with it (she related this dream to us, which we were unable to interpret); seven

[255] *Seeress*, p. 121.
[256] Ibid.
[257] Ibid., pp. 82–83.

> days afterwards Mrs. N– was delivered of a dead child, whose body was already in a state of corruption.[258]

And finally:

> On the night of the 28th January 1828, Mrs. H– dreamt that, being on a desert island, she saw her dead child enveloped with a heavenly light, with a wreath of flowers on its head, and a wand, with buds on it, in its hand. This disappeared; and she next saw me assisting a man who was bleeding; and this was succeeded by a third vision of herself, suffering severe spasms, whilst a voice told her that I was sent for. This dream she related to me on the morning of the 29th. On the 30th, I was sent for to a man who had been stabbed in the breast; and on the same night, the third vision was explained by my being sent for to her. The interpretation of the child's appearance we did not learn.[259]

These clairvoyant, prophetic dreams are related to her visions of ghosts. Of the appearance of these specters, she says:

> I see many with whom I come into no approximation, and others who come to me, with whom I converse, and who remain near me for months; I see them at various times by day and night, whether I am alone or in company. I am perfectly awake at the time, and am not sensible of any circumstance or sensation that calls them up. I see them alike whether I am strong or weak, plethoric or in a state of inanition, glad or sorrowful, amused, or otherwise; and I cannot dismiss them. Not that they are always with me, but they come at their own pleasure like mortal visiters [*sic*], and equally whether I am in a spiritual or corporeal state at the time. When I am in my calmest and most healthy sleep, they awaken me—I know not how, but I feel that I am awakened by them—and that I should have slept on had they not come to my bedside. I observe frequently that, when a ghost visits me by night, those who sleep in the same room with me are, by their dreams, made aware of its presence; they speak afterwards of the apparition they saw in their dream, although I have not breathed a syllable on the subject to them. Whilst the ghosts are with me, I see and hear everything

[258] Ibid., p. 83. Barbara Hannah notes that the "water in the dream stood for the amniotic fluid" (1959 [1934], p. 38).

[259] *Seeress*, pp. 83–84.

> around me as usual, and can think of other subjects; and though I can avert my eyes from them, it is difficult for me to do it—I feel in a sort of magnetic *rapport* with them. . . . If any object comes between me and them, they are hidden from me. I cannot see them with closed eyes, nor when I turn my face from them; but I am so sensible of their presence, that I could designate the exact spot they are standing upon. . . . Other persons who do not see them are frequently sensible of the effects of their proximity when they are with me; they have a disposition to faintness, and feel a constriction and oppression of the nerves. . . . The appearance of the ghosts is the same as when they were alive, but colourless. . . . Their gait is like the gait of the living, only that the better spirits seem to float, and the evil ones tread heavier; so that their footsteps may sometimes be heard, not by me alone, but by those who are with me. They have various ways of attracting attention by other sounds besides speech. . . . These sounds consist in sighing, knocking, noises as of the throwing of sand or gravel, rustling of paper, rolling of a ball, shuffling as in slippers, &c. &c.[260]

Such sounds are typical of spook stories, no matter in which country. I would like to read you another passage [*Jung cannot find the bookmark—to be read next time*].

Allow me to conclude today's lecture with some final remarks on this case: An intensive withdrawal from outer reality stimulates the background and the inner world, giving rise to three groups of phenomena: 1. extrasensory perceptions, such as clairvoyance, the perception of qualities through crystals, or perception through the epigastric region; 2. the apparition of ghosts; 3. the peculiar vision of the "sun-sphere" or mandala; "mandala" is the Indian term for "circle."

All so-called supernatural perceptions are mainly "clairvoyant" phenomena. Clairvoyance expresses itself through the senses and the spirit, and they in turn express themselves in terms of space and time.

We are at first inclined to regard the existence of such matters, which simply defy the laws of nature, as pure nonsense. But too much irrefutable factual experience has been obtained about them for them to be ignored. This should not amount to positing a metaphysical claim. We should merely exercise patience with such phenomena until we gradually discern what exactly is involved. Hardly a week passes in my

[260] Ibid., pp. 155–158.

practice in which a patient does not enter with such a dream or experience. Evidently, this field is filled with the most incredible possibilities of deception. Behind lurks a dark superstition, and yet our entire scientific world has emerged from precisely such a dark superstition, from a world of magic.

Lecture 9

15 December 1933

Submitted Questions

The first correspondent, a lady blessed with steady good fortune, is indignant that my lectures are so popular![261]

There are quite a number of reactions from younger members of the audience that have confirmed my worst fears. I would have spoken over the top of their heads, and they could not imagine for which reasons I have discussed at length such a curious case as that of the Seeress, which evidently dates from the last century![262]

I chose this case with a secret intention in mind. In so doing, I have espoused standard clinical methodology by selecting a classic example of an illness, offering a general description, and thereafter discussing the entire symptomatology and pathology of the illness based on the example. The case of the Seeress is an indisputably classic empirical example, and therefore allows us to consider certain basic facts. I have gone to some

[261] I.e., intelligible to the general public, in layman's terms. This refers to the following letter: "Zurich, 10 Dec. 1933. Dear Doctor, At our short encounter after your lecture about a fortnight ago you asked if your explanations would be popular enough. In the meantime, I have now heard from a number of people, students and people who have their feet on the ground, how they find your lectures. Surely you will be interested in learning what a small part of your large audience thinks. All of them find that the lectures are too popular. Not all of these people are specifically knowledgeable in psychology. Surely you will be relieved to know, since you are used to talking before a highly educated audience, that your very interested and attentive listeners can still follow you even if the basic concepts are explained at somewhat less length. In hoping that I could render you, dear Doctor, a small service with these lines, I remain, yours, Doris Schlumpf" (ETH Archives).

[262] This could refer to some reactions of students, gathered and summarized by a participant named Otto (ETH Archives; undated). They concur that the lectures did not meet their expectations, specifically, that the topics were too far-fetched and historical, and that Jung would not talk about contemporary problems and his own psychological theory. Jung also mentions similar complaints at the beginning of the thirteenth lecture; see pp. 106–107.

lengths to set out the details of the case to help you attain a clearer sense of the various phenomena involved.

Should the case strike you as unfamiliar or strange, then you would do so on account of your lack of knowledge. You are simply unaware that your own case exhibits all these basic facts, too, only they lie concealed in the dark background of your psyche. You have no knowledge of them, that is all. We must become better acquainted with some of the general features of the human psyche. This human psyche is nothing well-known, indeed it constitutes a great unknown. The ideas that I have set forth in my lectures on the basis of this case have already been published, and I am not to blame if these are not more widely known! For the moment, I shall not further discuss this. Bear with me as I proceed with my discussion of this case to a satisfactory conclusion, in order to alleviate for you some of the burden of misapprehension.

***

In the previous lecture, I highlighted the three characteristic phenomena:

1. extrasensory perceptions;
2. ghosts and specters;
3. the peculiar "sun-sphere."

With regard to these extrasensory perceptions, let us just assume that some of these curious facts about clairvoyance really apply. We must not let ourselves be deterred by superstition and fraudulent trickery. We can no longer ascertain the facts in the preceding case, but I have observed countless times that dreams and premonitions presaging the future do exist. We all know that, after all. One can even experiment with these matters, even in company, as I have done on countless occasions. They always happen whenever a person, like the Seeress, directs their entire attention inward, instead of outward. Then such dreams, premonitions, and perceptions occur that border on extrasensory perception. This is simply a fact, a quite uncomfortable fact, it is true, but nevertheless we must put up with it. It is my pleasure to admit as much. Unlike others who, for the sake of a theory, simply deny a whole bothersome part of a science and leave its treatment to the poets, I for one cannot allow myself that! If you so wish, I can give you my word of honor that such matters exist, and we can thus incorporate them in our conception. I am relating the case of the Seeress for precisely this reason. We thus face a most unpleasant and uncanny matter.

As a consequence, we must also defy the concepts of space and time. Of course, this relativization of time and space is unbearable for some mediocre brains, and is therefore simply denied. Such matters, however, do occur, wicked though this seems, and it is up to us to engage with them!

Already before our time, many of the brightest minds have found that, regarding time and space, things are not quite as they seem, that it is at least admissible to doubt the absoluteness of these dimensions. Kant, for instance, had serious doubts about these dimensions. He maintained:

> Space is a necessary representation, a priori, that is the ground of all outer intuitions. One can never represent that there is no space, though one can very well think that there are no objects to be encountered in it.

Space, he further asserts, is "a pure intuition," that is, "an *a priori* intuition"[263] that "grounds all concepts of it."[264] It has empirical reality; is the shape that all outer experience assumes.

Time, according to Kant, "is the *a priori* formal condition of all appearances in general."[265] In contrast to space as an external sense, time, as an internal sense, possesses "subjective reality."[266] Time in itself does not exist, because "time is nothing other than the form of inner sense, that is, of the intuition of our self and our inner state."[267] Sustaining a fundamental objection against this view will prove to be very difficult. Modern physics, as you are aware, has also come into conflict with these *a priori* concepts.

If time and space are relative dimensions, they cannot have absolute validity. Consequently, we must assume that an absolute reality has different properties from our spatial-temporal reality: in other words, there exists a space that is unlike our space, and a time that is unlike our time. That is, it is possible for phenomena to occur that are not subject to the conditions of time and space.

---

[263] Original: *Anschauung*. The editors and translators of the authoritative English edition, Paul Guyer and Allen W. Wood, translate this by "intuition," based on the fact that, in his (Latin) inaugural dissertation, "Kant uses the Latin word *intuitus* to signify the immediate and singular representations offered by the senses. . . . In the *Critique of Pure Reason*, he will employ the analogously formed German word *Anschauung* for the same purpose" (Kant, 1781/1787 [1998], p. 709).

[264] Ibid., A24, A25. Kant's book is usually quoted according to the so-called A/B system, "A" standing for the first, "B" for the second, revised and enlarged edition, followed, also in English texts, by the pagination in the standard German edition of Kant's works, the so-called *Akademie* edition.

[265] Ibid., A34.

[266] Ibid., A37.

[267] Ibid., A33.

Please bear in mind that psychic matters are neither thick nor thin, neither big nor small, neither round nor squared, neither heavy nor light, and so on, but exquisitely non-spatial; and secondly, that it is forbiddingly difficult to determine a time of the psyche. You will find it almost impossible to establish any time in which a psychic process occurs. You can measure response times, but what prevails is an enormously complex magnitude that consists of a whole array of quite unknown figures. In contrast, we have all had the strangest experience that under certain circumstances psychic processes require incredibly little time, for instance in so-called arousal dreams.

An example: a lengthy dream begins in times of peace; thereafter, we hear reports that war is imminent; heated debates for and against an armed conflict ensue, and the likelihood of hostilities intensifies. Newspapers report that deeds of war have been committed. Military personnel gathers, cannons are wheeled into position, suddenly heavy artillery guns begin to fire at rhythmic intervals, and eventually one awakes at someone knocking on the door.[268] Later, the dream can be described at length and in great detail, but in reality it lasted only for a very short time. Did this endless dream happen between the first knock and the last, then, or did it start earlier and lead up to the moment of the knocking from an anticipatory knowledge of the very second in which the knock at the door would occur? Of course, what the dreamer associates to such a dream is important, and the physician must pay attention to it. It is not necessarily the case that the contents of the dream were contingent upon the knocking on the door. One can also analyze such a dream, and discover that it springs from quite specific psychic conflicts.

Likewise, the fate of a poor soul being beheaded in a dream, whereby the decapitation was in actual fact a part of the four-poster bed crashing

[268] In October 1938, Jung used this and the following examples also in his keynote lecture on the method of dream interpretation during his seminar on children's dreams. There he revealed that his first example is in fact a dream he had had himself, and gave more details: "As a university student I had to get up at half past five in the morning, because the botany lecture started at seven o'clock. This was very tough for me. I always had to be awakened; the maid had to pound at the door until I finally woke up. So, once I had a very detailed dream. *I was reading the newspaper. It said that a certain tension between Switzerland and foreign countries had arisen. Then many people came and discussed the political situation; then there came another newspaper, and again it contained new telegrams and new articles. Many people got excited. Again there were discussions and scenes in the streets, and eventually mobilization: soldiers, artillery. Cannons were fired—now the war had broken out*—but it was the knocking on the door. I had the clear impression that the dream had lasted for a very long time and come to a climax with the knocking" (Jung, 1987 [2008], pp. 8–9).

down and striking the dreamer on the neck under the chin. It seems as if the dreamer had anticipated this event.[269] Or the well-known fact that people review their whole lives while falling down a mountain or while drowning, often reported by those who were saved. For instance, the admiral who took a false step.[270] Another case is Professor Heim, who once fell down a mountain and revisited his entire life during the fall.[271] Such cases seem to suggest that in certain circumstances the psyche needs only an unimaginably small amount of time.

We may now cite the positive evidence for the existence of nonspatiality, namely, when I can see through thick walls. In any event, we are touching upon a mode of existence that fails to coincide with empirically perceptible reality.

This point is most important with regard to the psychic being. I discussed the example of the "Seeress of Prevorst" for this very purpose—in order to show you how such great introversion results in the manifestation of the characteristic features of the psychic background, to the extent that the characteristics of consciousness completely vanish. These matters are by no means extraordinary, by the way. In theory, at least, each of you can have premonitory dreams. Generally, however, these are so insignificant that they are not noticed, but this is due to the ignorance of people. I myself was also so totally ignorant, but in the meantime I have known such cases for thirty years, and have published them, and if people don't know this it's certainly not my fault!

Another example is Dunne's *An Experiment with Time* (1927).[272] He was based in South Africa at a place where he received a mail delivery only every two months. On the eve of one particular delivery, he dreamt that he was reading a report about the disaster of St. Pelée in the *Daily*

---

[269] This is "an example from the French literature: Someone is dreaming: *He is in the French Revolution. He is persecuted and finally guillotined. He awakes when the blade is sliding down.* This is when a part of the frame of the canopy fell on his neck. So he must have dreamed the whole dream at the moment when the frame went down" (ibid., p. 8). Maury describes this dream of his and the incident in detail in the sixth chapter of his book on *Sleep and Dreams* (1861).

[270] "The same is told in the story of a French admiral. He fell into the water and nearly drowned. In this short moment, the images of his whole life passed before his eyes" (Jung, 1987 [2008], p. 9).

[271] Albert Heim (1849–1937), Swiss geologist, known for his studies of the Swiss Alps, professor of geology at the ETH (1873). "During the few seconds of his fall," he "saw his whole life in review" (ibid., p. 9).

[272] Dunne, 1927. John William Dunne (1875–1949) was an Anglo-Irish aeronautical engineer who pioneered the development of early military aircraft. After experiencing an anticipatory dream, he became strongly interested in parapsychology and the nature of time.

Mail or *Telegraph*, and was struck by a headline in bold print: "Disaster in Martinique destroys entire city, leaving 4,000 dead." The paper arrived the next day; he opened it eagerly and obviously it contained the report. But now to the most interesting aspect: Firstly, Dunne did not have his dream when the volcano erupted, and secondly he anticipated not the event but effectively a misprint. The next edition of the newspaper included the correction: 40,000 killed.[273] To Dunne, time is like a filmstrip, and the present is an observation slit; by mistake, it can happen that we merely look past the slit and see something which does not yet exist, that is to say, although it exists already as such, we cannot see it yet. Past events, too, can be seen in this way by looking past the time slit.[274] The actual soul, the objectively psychic, thus possesses qualities that border on nonspatiality and atemporality.

The second peculiarity is that the psychic background projects so-called ghosts. Naturally, we have no means at all of proving that ghosts exist. Like clairvoyance, this is a very complicated matter. At first, ghosts are no more than heterologous images of persons—with completely different faces and figures—quite often of persons who are no longer alive, whom one did not know, and whom the Romans called *imagines et lares*.[275] The Romans used the term *imago* to express the subjective nature of these images. The Seeress had to have all the souls of her ancestors within her in order to feel well. These are inner images, so-called autonomous

---

[273] See ibid., pp. 34–38. Jung here gives, in fact, a garbled rendition of Dunne's account. Dunne did not dream of reading a newspaper report, but of actually being on a volcanic mountain that was about to erupt and destroy the whole island on which it was standing, and with it its 4,000 unsuspecting inhabitants: "All through the dream the number of the people in danger obsessed my mind. I repeated it to every one I met, and, at the moment of waking, I was shouting to the 'Maire,' 'Listen! Four thousand people will be killed unless—'". When Dunne received the newspaper with the news of the catastrophe some time later (in his book he did not remember how long afterwards), the headline was: "Probable Loss of Over 40 000 [*sic*] Lives." He, however, *read* the number as 4,000, which number he also always quoted when telling his story later. Only when he copied out the paragraph fifteen years later did he realize that it was actually 40,000. It turned out, in addition, that the actual number of casualties, as found out later, did not coincide with that number. In his 1952 paper on synchronicity, Jung referred again to this episode, this time giving a correct version (§§ 852–853).

[274] This analogy with the filmstrip and the observation slit is Jung's, not Dunne's. The latter wrote, for instance, that "one had merely to arrest all obvious thinking of the past, and the future would become apparent in disconnected flashes," or that dreams in particular enjoy "a degree of temporal freedom" (1927, pp. 87, 164).

[275] Jung's own concept of the *imago* (what he later termed archetypal image) "has close parallels in . . . the ancient religious idea of the 'imagines et lares,'" that is, cult statues/images and tutelary Deities of home and hearth (*Transformations*, orig. Germ. ed. p. 164, also in *CW* 5, § 62 and note 5).

contents; they are autonomous, because these contents do not obey conscious intentions, but instead come and go as they please.

Now obviously you are bound to say: One does not have such things! Often, however, you yourself will say: "It has suddenly occurred to me," or such and such a thing has "just come into my mind." If you were Mrs. Hauffe, it would be a ghost addressing you. If you are slightly psychotic, then it is a voice that speaks behind you. "These thoughts have all been stolen from me, and now someone else is voicing them!" One man, for instance, used to hear a voice as loud and clear as a trumpet at nine o'clock in the evening; the voice would give an account of all his activities on that day. In another case, the voices read aloud all the company nameplates on this person's way home in London. When something fails to work in our psychology, all these matters surface—matters that one does not believe one has perceived.

Once a patient was brought to me in a highly neurotic state, an eighteen-year-old girl who had enjoyed the best education, and led an extremely sheltered life. To the shock of her parents, when she was agitated she would utter a flood of the most incredible expletives, on which even a wagoner could have prided himself. "Could you please explain how on earth this child knows such foul-mouthed language!" Now I couldn't tell them where exactly the girl had picked up those expressions, but the fact is that she had actually heard them, be it from a cabman on the street, be it from other kids, etc. We have a very high threshold of consciousness. Our consciousness can be aroused only by phenomena that possess sufficient energy; everything else does not become conscious, although it is perceived. It is like sound vibrations. Do you believe that sound ceases when you no longer hear it? If the light were suddenly to go out and you could no longer see me, you would not be likely to think that I had ceased to exist, yet it would be no more foolish to think so than to assume that the contents of the psychic background only exist when we can see them.

These phenomena point to the actual quality of the psyche. We learn from them that autonomous contents are one of the essential facts of the soul; that is, they are independent contents that abide by their own laws, and that come and go and produce characteristic moods. This is well expressed in our language. We say, for instance: "What has got into him again?" Or that someone is "possessed"—by a spirit, that is, or that he is "beside himself." We also talk of "jumping out of one's skin" or that someone is "bedeviled" by something.

The ancients understood this far better than we do; they did not speak, therefore, of being in love, but of being possessed or hit by a God. We not only experience these psychic contents as a state of possession, but also as a sense of loss, for the unconscious can steal away fragments of our conscious psyche and rob us of our energy. This is what happens when we say that we would not be "in the mood" for something or other. The primitive would say that a spirit—possibly one of the ancestral souls whose presence he must sense to feel at ease—"has gone forth" and left him. The souls of the ancestors—these are autonomous contents, hereditary contents, of which your soul consists. Is it not true that you have your grandmother's nose, and so forth? It happens that one will suddenly hear a terrible din in a native village. A Negro is lying on the ground, beating himself. It is clear to everybody that a spirit has left him, a spirit he cannot do without, however, and that he tries to call back. "His soul has gone astray." It is as if he were seeking to remember himself, thereby causing him to inflict pain upon himself. Quite like when one grows restless in a boring lecture.

We cannot escape being influenced by psychic contents, it is our natural condition. Therefore I always feel very suspicious when somebody assures me that he is completely normal. It is well established, however, that very "normal" people are compensated madmen. Normality is always slightly suspicious. I'm not just joking, but this has been the bitterest experience of my life.[276] Generally, it suddenly becomes evident that some madness lies concealed beneath their cursed normality. The truly normal person has no need to be always correct, or to stress his normality. He is full of mistakes, commits follies, lacks modesty, and does not hold normal views.

When someone runs around strangely among primitives, eating only grass and so forth, he is said to be "possessed," that is, by the devil. "Here we go again," is the response, or one would like to "jump out of one's skin," if only one could. It is a like a soul jumping out of the skin. We are "possessed" when a soul enters, and "jump out of our skin" when a soul leaves us. It is these autonomous contents that lead to so-called possession. Even the most normal person can be possessed, by an idea, for instance, or a conviction, or an affect.

Now to the third phenomenon, the peculiar circle. This is the most curious phenomenon of all. Unfortunately, this fact is completely unknown. I fear, however, that you have to grow accustomed to these matters, even to those you know nothing about yet. I exercise caution

[276] Sidler made a note: "Meaning: When Jung thinks to have finally found a 'normal' follower or pupil it turns out that some kind of madness lurks behind his cursed normality."

in these matters, and have therefore chosen a case in which I was not involved in the least; otherwise, one would say again: "Well, of course, he simply influenced the patient's mind!" This is about a basic fact, one which has received merely scant attention to date. It is nothing other than an absolutely basic fact about the human soul; it is known all over the world and, if we do not know it, then we are the morons![277]

I myself have witnessed a case where this phenomenon occurred. This was a girl aged sixteen who exhibited this phenomenon whom I observed almost thirty-seven years ago, in the final semesters of my university studies. I was completely ignorant of such matters at the time. She had drawn a circle on the basis of information that she had received from spirits. . . .[278]

These reflections bring us to an instance of primitive psychology that is still apparent today—although perhaps not so much in this auditorium, but by all means with local councilors of various Swiss municipalities.

The next analogue of this circle is the so-called magic circle: *Doctor Faust's Coercion of Hell*,[279] for instance, contains recipes for the invocation of ghosts. We have lost all knowledge pertaining to these matters. There are three circles that serve to ward off evil spirits rather than evil persons.

---

[277] Here all extant lecture notes, with the exception of Sidler's, break off. As we know from a letter of a participant, Arthur Curti, Jung overran his allotted time past 7 p.m., and the majority of the audience left or had to leave at this point (letter of 19 January 1934; ETH Archives; see the beginning of lecture 11 and note 297). Sidler's notes on the rest of the lecture are very sketchy, however, sometimes to the point of unintelligibility.

[278] At the time (that is, around 1895/1896), Jung experimented with his cousin Helene Preiswerk, the person he is referring to here, and who was the subject of his doctoral dissertation (1902). Sidler's notes on Jung's presentation of these circles are so fragmentary and obscure that they have been omitted here. The reader will find a reproduction as well as a detailed description and discussion of these circles in ibid., §§ 65 sqq.

[279] Dr. Johann Georg Faust (ca. 1480–ca. 1540), itinerant alchemist, astrologer, and magician, a legendary figure about whose life hardly any facts are known with certainty. He became the model for various literary works on the Faust theme, among them Goethe's play. Various so-called *Höllenzwänge* (e.g., Faust, 1501) are ascribed to him. A *Höllenzwang* ("Coercion of Hell," or "Hell-Master") is the designation or the title of grimoires, or spellbooks, that describe rites or incantations with which the demons of Hell can be forced to obey the commands of the magician.

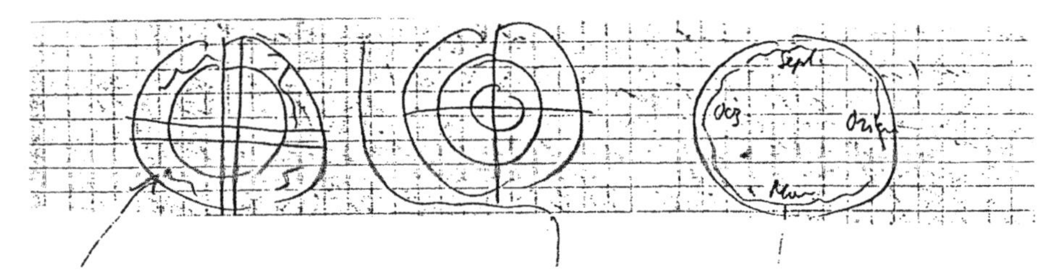

The first circle exhibits strict correspondence with the form of a cross; it bears inscriptions of names, as a rule the names of Hebrew Gods. In the second circle, the names of the Gods serve as a protective barrier against evil spirits. The third one also contains the names of Gods, arranged in circular fashion. Another popular form is the following:

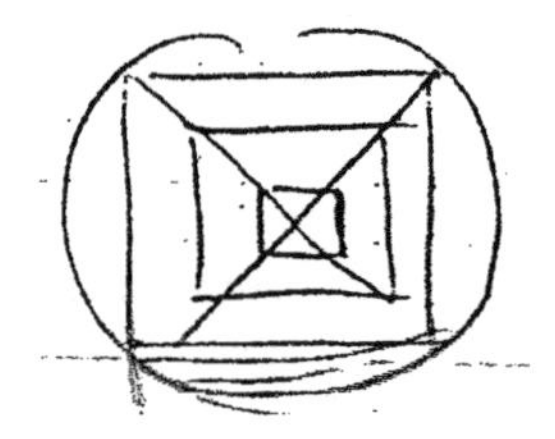

Usually, the invocation of a ghost proceeds from the center, involving the recitation of the names of all Gods, of both the conscious and secret ones, of the four winds, and of the [*sic*]. Ordinarily, furthermore, the invocation is spoken at all four points of the compass rose: "I (my name) herewith bless and consecrate this circle in the names of the highest God. Ego . . . [*sic*][280] who stand in this circle, so that the Almightiest God may bestow upon me and all others a shield and a defense against all evil spirits and their powers, in the name of God the Father +, the Son +, and the Holy Ghost +." This simple onomastic formula is used to afford the circle magic potency.

This circle sprang from ancient customs, which are still practiced. For example, circumambulation: the ritualistic practice of circling on foot, clockwise and three times, whatever is to be banished or protected. When the Romans founded a city, they encircled the *sulcus primigenius* [original furrow]. A pit, known as the *fundus*, was excavated at the center; the temple, in which all kinds of objects were deposited, was built there.[281]

---

[280] The notetaker did not catch all of this phrase.

[281] Jung alluded briefly to this tradition in his commentary on *The Secret of the Golden Flower* (1929, § 36), but gave the most detailed description in his seminar on *Dream*

The oldest depictions of circles, so-called sun wheels, date to the Paleolithic period.[282] Please note that wheels did not yet exist at the time; the first wheels appeared in the "Wooden Age."[283] The oldest "sun wheel" is an octagonal cross, to which Frobenius already refers.[284] Also at the Swiss National Museum [*Landesmuseum*] in Zürich.

Encircling a municipality or property on horseback is an old magic custom serving its protection against evil spirits.[285] Sometimes, (chickens?) [*sic*] are (tended?) [*sic*] behind the so-called blocking chain [*Sperrkette*] about once a year: that is, a magic circle is drawn behind the chain beyond which the animals are not allowed to roam. Evidence from all over the world attests that the central cultic notions are furnished with this symbol.

There is a Chinese manala, in which cosmological forces emanate from the four ends of the cross.

In Egypt we find the same idea. The sun is positioned in the center, and around it are the four sons of Horus in quadratic order. Only one of these four figures has a human head, the other three have animal heads. This corresponds to the Christian symbol of the tetramorph—the cross with the four Evangelists, of whom only one has a human head.[286]

Interpretation Ancient & Modern (2014, p. 213): "Jung: What was done when the [ancient Roman] city was founded? Participant: One walked around the *temenos* [a protective piece of land set apart as a sacred domain, a sacred precinct or temple enclosure, set off and dedicated to a God]. Jung: How was that done? Participant: By circumambulations. Jung: Yes, and with what? Participant: With a plow. Jung: Yes, it was used to plow the *sulcus primigenius* [a magical furrow around the center of the temple *temenos*]. This was a mandala; and what was done in the middle of this plowed up area? Participant: Fruits and sacrifices were buried. Jung: At first a hole, the *fundus*, was made, and then sacrifices to the chthonic Gods were put into it; in other words, the center was accentuated." Jung seems to allude to the older Roman method of surveying field boundaries and building sites described by Pliny in the *Natural History* (Book III). Cf. Rykwert, 1988.

[282] I.e., the earliest period of the Stone Age, lasting until circa 8,000 BCE.

[283] Jung's neologism.

[284] Leo Viktor Frobenius (1873–1938), German ethnologist, archaeologist, and explorer, and a leading expert on prehistoric art and culture. He organized twelve expeditions to Africa between 1904 and 1935. In 1920, he founded the *Institute for Cultural Morphology* in Munich. Frobenius, and in particular his book, *Das Zeitalter des Sonnengottes* [*The Age of the Sun-God*] (1904), was a major source of reference for Jung, above all in *Transformations*.

[285] Jung may also have had in mind the so-called *Bannumritt*, a tradition to this day in rural areas around his hometown Basel. In circling their communities, the inhabitants fire from old muskets.

[286] Jung repeatedly pointed out the connections between Horus and Christ, and the depictions of the sons of Horus and the Christian tetramorph; see in particular Jung, 1945a, §§ 360 sqq.; 1951, §§ 187 sqq.

The Mayan "Temple of the Warriors" was excavated a few years ago. Beneath the altar a mandala, consisting entirely of cut turquoises, was found encased in a limestone cylinder. It was bedecked with 3,000 turquoises. It is kept at the Museum of Mexico City.[287] In the four main points comes the feathered serpent[288] and opens its mouth inward. This serpent adorns the robes worn by priests to this day, and it has a spellbinding effect in that whoever looks at it is enchanted. [It is] The object of concentration. Whoever succeeds in placing themselves in this circle is protected against evil spirits.

In India:[289]
In Tibet:
right inside: a precious object, a symbol of—or the image of the highest Goddess.
Circle: a pagoda with four entrances
Paramahansa, see Paul Deussen p. 703, *Zentralbibliothek* [main library]. A conversation found in the Para[mahansa] Upanishads:[290]

The Pupil asks: At whose wish does the mind sent forth proceed on its errand? At whose command does the

[287] In his seminar on dream analysis, in the session of 13 February 1929, Jung gave a more detailed description of this discovery (actually not an excavation) at Chichén Itzá, closely following the account published in the *Illustrated London News* of 26 January 1929, which he had evidently read: "An American explorer has broken through the outer wall of the pyramid and discovered it was not the original temple; a much older, smaller one was inside it. The space between the two was filled with rubbish, and when he cleared this out he came to the walls of the older temple. Because he knew that it had been the custom to bury ritual treasure under the floor as a sort of a charm, he dug up the floor of the terrace and found a cylindrical limestone jar about a foot high. When he lifted the lid he found inside a wooden plate on which was fixed a mosaic design. It was a mandala based on the principle of eight, a circle inside of green and turquoise-blue fields. These fields were filled with reptile heads, lizard claws, etc." (Jung, 1984, p. 115). The inner temple is called the Temple of the Chac Mool. The archeological expedition and restoration of this building were carried out by the Carnegie Institute of Washington from 1925–1928. A key member of this restoration was Earl H. Morris who also published the work from this expedition (1931).

[288] The feathered serpent was a major deity of the ancient Mexican pantheon.

[289] The following section is incomplete in the notes.

[290] Deussen's translation: *Sechzig Upanishad's des Veda* (1897; 2nd ed. 1905, 3rd ed. 1921).—The Paramahansa Upanishad describes the conversation between Sage Narada and Lord Brahma about the characteristics of a Paramahansa. The latter is a Sanskrit religio-theological title of honor applied to Hindu spiritual teachers of high status who are regarded as having attained enlightenment. The following fragmentary notes, however, do *not* refer to the Paramahansa Upanishad, but to the Kena Upanishad, which Sidler then probably copied in full from the Deussen translation only after the lecture and attached it to his notes.

first breath go forth? At whose wish do we utter this speech? What God directs the eye, or the ear?

The Teacher: It is the ear of the ear, the mind of the mind, the speech of speech, the breath of breath, and the eye of the eye.

[*sic*] is not expressed by speech and by which speech is expressed, that alone know as Brahman, not that which people here adore.

The Kena Upanishad of Sâmaveda (its older name is Talavâkara Upanishad, since it originally belonged to either the Brâhmana estate of the Talavakâra or Saiminêya).[291]

First Khanda

1. The Pupil asks: "At whose wish does the mind sent forth proceed on its errand? At whose command does the first breath go forth? At whose wish do we utter this speech? What God directs the eye, or the ear?"
2. The Teacher replies: "It is the ear of the ear, the mind of the mind, the speech of speech, the breath of breath, and the eye of the eye. When freed (from the senses) the wise, on departing from this world, become immortal.
3. The eye does not go thither, nor speech, nor mind. We do not know, we do not understand, how any one can teach it.
4. It is different from the known, it is also above the unknown, thus we have heard from those of old, who taught us this.
5. That which is not expressed by speech and by which speech is expressed, that alone know as Brahman, not that which people here adore.
6. That which does not think by mind, and by which, they say, mind is thought, that alone know as Brahman, not that which people here adore.
7. That which does not see by the eye, and by which one sees (the work of) the eyes, that alone know as Brahman, not that which people here adore.
8. That which does not hear by the ear, and by which the ear is heard, that alone know as Brahman, not that which people here adore.

[291] This explanation was obviously added by Sidler.

9. That which does not breathe by breath, and by which breath is drawn, that alone know as Brahman, not that which people here adore."

Second Khanda

1. The Teacher says: "If thou thinkest I know it well, then thou knowest surely but little, what is that form of Brahman known, it may be, to thee?"
2. The Pupil says: "I do not think I know it well, nor do I know that I do not know it. He among us who knows this, he knows it, nor does he know that he does not know it.
3. He by whom it (Brahman) is not thought, by him it is thought; he by whom it is thought, knows it not. It is not understood by those who understand it, it is understood by those who do not understand it.
4. It is thought to be known (as if) by awakening, and (then) we obtain immortality indeed. By the Self we obtain strength, by knowledge we obtain immortality.
5. If a man know this here, that is the true (end of life); if he does not know this here, then there is great destruction (new births). The wise who have thought on all things (and recognized the Self in them) become immortal, when they have departed from this world."[292]

[292] There are many different translations of the Kena Upanishad into English. I have chosen the one at http://www.tititudorancea.org/z/kenaupanishad_talavakara_1.htm.

Lecture 10

12 January 1934[293]

In the last lecture, we concluded our discussion of the "Seeress of Prevorst." Her case illustrates the utmost introversion, in that everything moves inward. Everything in her has been cut off from our reality; indeed she defended herself against the outer world. To her, reality as we know it has been dispossessed of its meaning and emotional emphasis; instead, something unknown to us, about which we know only through legend, appears in her. To her, this background of the soul is plastic, and possesses all the meaning and emotional emphasis that reality has for us. Whereas human beings—some loved, others loathed—inhabit our world, ghosts populate hers. Where the sun or the moon shine in our world, in hers shine an inner sun and an inner moon.

Wherever and in whomever we find such pronounced introversion, we will also find indications of these phenomena. If we ask those people, they will deny it, and for several reasons. First, because they are shy of exposing themselves to ridicule by admitting to such experience; the Seeress, however, was too deeply convinced of the reality of her experiences to be troubled by such fears. Second, people are as a rule afraid of these things, for they have heard that they belong to the field of psychiatry. Third, because they are very often unconscious of these experiences, and as a consequence suffer indirectly from symptoms.

Whenever introversion intensifies, the three phenomena I mentioned become apparent:

1. Time and space become relative, presentiments and dreams come true, and telepathic experience occurs.
2. We find certain autonomous psychic contents, ultimately leading to personifications and the apparition of ghosts.

[293] Sidler noted that he had to miss this lecture due to tonsillitis, so the compiled text of this lecture is based on the notes by M. J. Schmid, R. Schärf, and B. Hannah.

3. Symbols of a psychic center are experienced. This center does not coincide with consciousness, and is generally perceived as a source of life, equivalent to an experience of God. One can recognize therein the essence of religion.

The Seeress is most certainly a border case. While it is very rare to encounter such cases, quite a number of them have been recorded throughout history. In contrast, cases involving compensation are more frequent. These cases will not strike us as strange as the one we discussed, and we will see more clearly how familiar we all are with such experiences. If people are not destined to die in a state of complete introversion, a reaction will have set in, and a certain extraversion will become apparent. The background of the soul is clouded over, the energetic charge of the contents decreases, plasticity wanes, and the images become pallid and blurred. We find all indications of an outer reality that interferes with the background. The image of the background of the soul becomes translated into the banality of everyday life, and the spotlight of experience is directed elsewhere. We will consider the main stages of this process, not in regard to any particular case, but as I have been able to observe it generally.

In the first phase, the centerpoint vanishes, just as the sun-sphere did in the case of the Seeress. This vision of the sun grows dim. While it might remain intuited, it ceases to play a role. In many cases, it becomes unconscious. What remains is the intermediate realm of so-called ghosts and of those phenomena that manifest some uncertainty regarding time and space, that is, experiences of a telepathic nature.

In the second phase, the autonomous figures, namely, the personifications and ghosts, disappear. Presentiments and telepathic dreams continue to exist, as well as curious manifestations in consciousness that elude rational explanation. What is known appears strange; forgetfulness occurs, partial amnesia, and so forth. Only vestiges remain of the autonomous contents, which had previously become personified. Such phenomena can still be observed in primitives, who ascribe them to the presence of ghosts. It is always a ghost, a witch, or a sorcerer who takes something away from them. With the mentally ill, this condition is known as "thought withdrawal."

The third phase occurs when the entire psychic background goes dark, that is, when nothing remains of these autonomous inner psychic phenomena. The person's memory seems to be normal, and the psychic [autonomous inner] phenomena no longer seem to exist. Here we approach "normality." The more manifest so-called normality becomes, the more a strange phenomenon occurs, namely a defensive attitude towards the matters of the background that no longer appear attractive. One is no longer tempted to have dreams and to experience the appearance of ghosts. The entire affair becomes uncanny, repugnant, repulsive, childish, ridiculous. Such people begin to build a thick wall of rationalist skepticism and "scientific" attitudes, and seal the entire matter airtight. If anything creeps through nonetheless, it is dismissed as "merely psychological." This, however, prompts a true and proper witches' sabbath of incompatible complexes. Consciousness grows too strong in proportion to the degree in which the background is walled off. Those people find themselves terribly interesting and important, and become most dreadful bores. It is nothing but an exaggeration, a showing off, but in a manner that has lost the characteristics of experience; it is a pumped-up story, something that is intended to impress. If such a condition persists, we speak of neurosis. An inflation of the subject occurs, and everything becomes psychologized. Since those people are wrong and actually know that they are, however, they become oversensitive—hypersensitivity is always suspicious!—and one has to walk on eggshells around them in order not to tread on their psychological toes.

This is an unfortunate intermediate condition that improves immediately, however, when extraversion actually commences, and all thought of the wall and world beyond is forgotten and obscured. Neurosis will then abate. Such a man no longer looks into himself, but turns to the conscious world with a sense of relief and freedom, and thus detaches himself from the background. His friends will push him still further along that path: he should meet people, he should travel, throw himself into something, he shouldn't waste his time, but exercise his will, etc. Such people become veritable acrobats of the will. So-called objective values become increasingly persuasive, and it is extremely important to them to be normal and healthy. Such concepts are indeed effective—the concept of "normality" happens to be persuasive, although no one knows what "normal" means. The inner world is now completely darkened, and appears only here and there in the form of slight disturbances. "I feel absolutely terrific. I'm always happy, satisfied, etc." Such an attitude of "healthy-mindedness"[294] is typical of Americans, and is based entirely on the extraverted principle. All goes swimmingly, he overflows with wonderful descriptions of his

[294] This expression, probably referring to William James's concept of "healthy-mindedness," is in English in the German lecture notes.

family and his enviable lot—till one day he appears with a face a yard long because he has had a bad dream, and something from behind the Great Wall has managed to slip through. Dreams are incursions from the hinterland, and the shadow announces itself. The closer people still are to that Great Wall, the better they hear what happens there—but then they immediately rationalize it again.

A patient once consulted me in precisely such a state of agitated extraversion. I advised him to spend an hour on his own every day. He was delighted at the thought, and said that he could now play the piano with his wife every day, or read, or write. But when I discarded each of these possibilities one after the other, and explained that he was to be really alone, he looked at me in despair and exclaimed: "But then I'd become melancholic!"

The next phase is complete extraversion. Here, we find the people who have already become quite identical with what they represent, and no longer with what they are. We often see this in people who have been successful, for example, the village or town mayor is nothing other than that. Such achievers live their roles day and night; they are already living their biographies as it were. They appear solemn, and radiate a persuasive dignity and conformity; everything is well balanced. It is actually the highest level of perfection on this way if someone becomes identical with the object, that is, with the light in which he wishes to appear. He doesn't have a clue about his own subject, but has become completely absorbed in something else, and is no longer himself. He has devoted himself to something, and has now become that thing, his status, his profession, or his business. Individuals who thrive completely on an object have subjective motives.

There is a good story about a Basel parson which illustrates this condition admirably—psychology consists of good stories! He was full of zeal for the welfare of his congregation and eager to provide it with the recreation he felt it required, but he was poor—such people choose their parents badly and never have any money, they always have to beg it from others! On his rounds among the richer Basel citizens, he called on a very sarcastic professor of theology who was well furnished with this world's goods. After much pleading, during which the professor remained unmoved, the parson leapt up in a rage, screaming: "*D'r* Herr *will's!*" [The *Lord* wants it!] The professor, pointing at him, replied: "Der *Herr will's!*" [*This* gentleman wants it]. This road leads to the illusion that what I, a lamentably small I, want, is the will of God.

This outward movement, however, is not entirely ridiculous, but is part of the birth process of man. Children and adolescents *must* forget the background. A child who remembers the background for too long would become inept at entering the world. Young people *must* erect many walls between the background and the subject so that they can believe in the world. Otherwise we would be unable to do anything. If we are anchored in the inner world, the values of the world will seem doubtful, and we will be unable to pull ourselves together for some real action. We will get lost in thoughts, and miss the right moments, or fail to attach to a given matter the vital importance it actually has. Many become procrastinators because they are anchored too much in the background, and can no longer summon faith in what matters. To be wholly devoted to something is also an art and something good, in particular for young people.

For young people, this is absolutely important. It is true, however, that there are also young people who are philosophically or religiously minded, and those should indeed know that something else exists, too, and that one should not live only outward. For if they misplace the values that belong in the background and move them to foreground, their world view will become distorted. Many difficulties arise from the fact that relationships and values are treated with an importance they do not deserve. For instance, Mr. So-and-so said this and that about me. I could take terrible offence. This, however, would be ridiculous, and not worth dwelling upon. For what have we said about others!

I would like to illustrate this outward development with a diagram.[295] The right side illustrates how this development unfolds in the outer world, as I have discussed in this lecture. The centerpoint is the subject, to which everything else refers.

[295] Hannah places this passage and the diagram at the beginning of the next lecture, dated 19 January 1934.

RIGHT | LEFT

HIGHEST IDEALS	ABSOLUTE IDEAS	PEOPLE	PROJECTIONS PRESUPPOSITIONS	COMPLEXES	PROJECTIONS	SPIRITS GHOSTS	ABSCON-DITUS
Godhead Sun	Pope King	NON EGO	Impressions	Affects Hunches Ideas Dreams	NON EGO		Mystical Godhead Circle
	State Church Party	Time Space Causality	Perception through the senses			Demons Angels Symbols	
	Abstract	Objective	Subjective Ego centric Auto erotic		Objective	Abstract	
IV +	III +	II +	I +	I –	II –	III –	IV –

SUBJECT

Fringe of consciousness

Lecture 11

19 January 1934[296]

Submitted Questions

There is a query from one gentleman regarding the mandalas discussed at the end of the last lecture. I shall explain it to him after today's lecture.[297]

***

Today, I would first like to make some further remarks on the diagram that I presented last week.

Right Side of the Diagram

The impressions of our surroundings that we perceive are not simply the objects themselves, but what we perceive are the images of these objects and persons. We perceive things as they appear to us, and we are always caught in subjective prejudices that distort and disturb our perceptions. For instance, "Mr. X makes an excellent impression on me—but not on my friend." Or "I find the painting very nice, but my friend thinks it's dreadful." It is as if we were surrounded by an oddly deceptive fog

[296] R. Schärf's typescript of this lecture is missing.

[297] "Zurich, 19 January 1934

Dear Doctor!

At the end of the *next-to*-last lecture you quickly went over the contents of the various mandalas. As it went so fast, however, I could not keep up with my note taking, and my notes on the descriptions of the ancient *Maya* and *Indian* mandalas, as well as of the conversation between the pupil and the master (*Upanishads*), are incomplete. I would be very grateful, therefore, if you could again *briefly* describe the *contents* of these mandalas, perhaps also that of the *Egyptian* mandala, at the beginning of the next lecture.

Since the majority of your listeners did not hear your explanations at all, because they had to leave after 7 p.m., I think this would be of interest to others, too.

With many thanks and best wishes,
Arthur Curti" (ETH Archives).

through which we perceive things, and which affects our perception in a peculiar way. No human can escape this. It is the greatest possible art to observe objectively. We probably observe rather precisely things that are indifferent to us, but in the case of things that concern us personally the fog gets thicker.

Why is this so? We have unconscious presuppositions or false associations. A female patient once told me: "My doctor strongly objects to my consulting you." "How so, I don't even know this gentleman. What did he tell you?" "I don't know." "Well, did he tell you directly?" "No, but I could tell." "How so?" "He said: 'What's this, you intend to consult that mad baldhead?!'" Clearly, the man must have had an utterly wrong notion of me when he called me baldheaded![298] William James refers to the fog that surrounds us as the "fringe of consciousness."[299]

The further things lie from us, the more objective we can be. In the sphere of abstract ideas (III), an impersonal or not-I way of looking at things exists, which is quite free from subjective prejudice. It is impossible to live entirely in the personal attitude, because the nonpersonal catches us somehow. We need both the personal and the impersonal point of view. To approach the Divinity has always been felt as an escape from the futility of personal existence. I once saw something very touching in the newly excavated tomb of a pharaoh: a little basket made of reeds stood in a corner and in it lay the body of a baby. A workman had evidently slipped it in at the last minute before the tomb was sealed up. He himself was living out his life of drudgery, but he hoped that his child would climb with the Pharaoh into the ship of Ra and reach the sun.[300]

But the personal element is also necessary in life. A woman once came to me absolutely broken down because her dog had died. She had drifted away from all human contact, the dog was her only relationship; when it disappeared, she went to pieces. The primitive makes no distinction between the personal and impersonal. *L'état c'est moi*, as Louis XIV said,[301] and that is just how the primitive king looks upon his kingdom.

Nature simply produces something; she never tells us her laws. It is human intelligence that discovers them and makes abstractions and

[298] Sidler notes here: "As if by accident, Jung turned the back of his head towards us to reveal his bald patch."

[299] "Let us use the words *psychic overtone*, *suffusion*, or *fringe*, to designate the influence of a faint brain-process upon our thought, as it makes it aware of relations and objects but dimly perceived" (James, 1890a, p. 258; on "fringe of consciousness," see p. 686).

[300] Jung tells the same story (although in a slightly different version) in 1939b, § 239.

[301] "I am the state," the maxim of absolutism, probably falsely attributed to Louis XIV of France (1638–1715).

classifications. In the abstract sphere (III), things and persons are classified by their sex, age, family, tribe, race, people, language, and settlement area; and thereafter by occupation, and by psychological and anthropological types. These are so-called natural classes, which, however, also correspond to an abstraction, so they belong to Section III + in the diagram. Abstractions can become more important than the human unit; it is a question here of "how many?", not of "who?"

The abstract sphere also contains purely ideational groups, which are characterized by a particular idea, such as the state, the Church, political parties, religions, various -isms, societies, and so forth. They usually possess a symbol, for example, the ideational group of the Swiss fatherland is symbolized by a white cross against a red background, or take the cross as the Christian symbol, the half-moon, the Soviet star, and the swastika. Or the totem animals of the nations, such as the Prussian eagle, the British lion, or the Gallic cock. Kings wear the crown, the corona, and the coronation mantle, the astral sphere. The old German emperor holds a *globus cruciger* in his hand, the symbol of earthly power. These are all astral, cosmic symbols. These ideational groups deny any purely natural descent, but instead claim to stem from the most exclusive and oldest ancestor: the sun. Symbols such as the sun symbol belong to the sphere of highest ideals (IV +); the sun often symbolizes the father, the life giver. This brings us to the end of the right side of the diagram, the side of consciousness.

Left Side of the Diagram

But how are things on the other side, the backside? We have already discovered that we have a shadow side, too, a "behind." This is something that we leave in the shadows, that we prefer to leave untouched, except perhaps under particular circumstances, such as when we go to confession, or when someone "professes" his allegiance to the Salvation Army.[302]

Certain contents may surface. You might, for instance, awaken in a terrible mood: "What's wrong with you?" "Nothing at all!" Within the "fringe of consciousness" lies the province of projections, of affects, and of inexplicable moods. We are also blessed with ideas or, if you want to

[302] A Salvationist Soldier has to sign the *Salvation Army Articles of War*, expressing his commitment to live for God, to abstain from alcohol, tobacco, and from anything that could make his body, soul, or mind addicted, as well as expressing his commitment to the ministry and work of a local Salvation Army corps.

put it more nobly, with inspirations. The Americans have a good word in this respect: to have a hunch,[303] that is, a humped or crooked position. And, obviously we have dreams, too.

Let us now proceed somewhat further. Just let us assume, as a working hypothesis, that things in this sphere of unconscious presuppositions are not quite the same as on the other side. It is not likely that ideas spring from the aether, but they are based on underlying psychological material. Likewise dreams: in analyzing them, we discover all kinds of material that was previously unconscious. Let us make another hypothesis, namely, that these unconscious presuppositions distort affects and ideas in the same way in which the fog distorts perceptions on the other side.

Now something analogous ought to correspond to people and things on the other side, and these are the phantoms and ghosts. It is the ghosts that cause those affects, ideas, or dreams, and they might also explain why we awaken in a certain mood. This is really the case in primitives and the mentally ill. The primitive has a better realization of the autonomy of this inner side than we have. He does not speak of *having* a mood, but of being *possessed* by one. So he does not say, "I am angry (or sad)," but instead "I have been made angry (or sad)," or "I have been shot at." Likewise, the mentally ill exclaim, "X-rays shot this through me." "They," the spirits or ghosts, steal the primitive's soul away and make him ill, so he knows that he has to work day and night to remain aware of them and keep them at bay. There is a native tribe in Australia that, even when temperatures drop to two degrees Celsius below zero, will lie naked beside the cold fire, freezing stiff. But they would never dream of covering themselves with furs and blankets! One tribe actually spends two-thirds of its time fighting the magic influences of evil demons! But you try telling any of them that this is nonsense! I would advise you not to! I once tried—and my successor, who also tried, was spiked with spears!

We saw that the Seeress's world was peopled with ghosts, just as the outer world is inhabited by real people and objects. So the spirit world is the complete equivalent of the outer world on the other side of the diagram. For us, this sounds like a crazy story; this is something that leads us deep into the primordial world. Like the people in the outer world, ghosts form groups, too. For example, the Church organizes its angels in a celestial hierarchy of nine orders and three groups, this hierarchy reaching its zenith in the Godhead.[304] Two hundred years ago, the world

---

[303] This expression in English in the German notes.

[304] The Catholic Church distinguishes between nine levels, or choirs, of angels (Seraphim, Cherubim, Thrones, Dominions, Virtues, Powers, Principalities, Archangels, and

of demons was still alive for us. For Paracelsus, this was still a world of realities. But also for many people living in our mountains, these things are still real. So when a mountain farmer notices one day that his cow is giving less milk than usual, he immediately runs to the Capuchin monk to fetch a prayer card of Saint Anthony.[305] The next day, the cow might in fact give again as much milk as before. The farmer, however, tells no one that he has been to see the monk; on the contrary, should anyone inquire whether he believes in demons or ghosts, he will laugh and exclaim: "Oh no, what nonsense!" But he says this only because he is seeking election to the local council.

With primitives, things are quite different. I discussed these matters at length with a tribe in Central Africa that I once stayed with.[306] Our discussion afforded me highly interesting insights, which the relevant literature confirmed also for many other peoples. For here we are at once confronted with a peculiar problem, because in primitives the center, the I, is obviously missing; in its place there is a plurality, a multiplicity. This is due to the unconscious identity, the *participation mystique*, of primitives with each other and with objects.[307] The tribe comes to stand for the "I." This accounts for the strange nature of their customs, such as the following: A horse has been stolen. The medicine man summons all male members of the tribe, has them stand in a circle, sniffs at each of them, and ultimately says to one of them that he is the thief. The latter acquiesces to this procedure without protest, and is put to death. Because being a member of the same tribe he, too, could have been the thief, even if was innocent in this particular instance.

Their own personal life means very little to them, a native will even commit suicide in order that his ghost may haunt the thief who has

---

Angels), organized in three triads, or spheres.

[305] This is known as Saint Anthony's "Brief" or "Letter," which he allegedly gave to a woman who wanted to drown herself, thereby freeing her from demonic oppression and the desire to do away with herself. It contains the following exorcism: *Ecce crucem Domini! Fugite partes adversae! Vicit Leo de tribu Juda, radix David. Alleluia!* [Behold the Cross of the Lord! Flee ye adversaries! The Lion of the Tribe of Juda, the Root [son] of David, has conquered. Halleluia!]

[306] The Elgonyi in Kenya, in 1925/1926. See Jung's description of this trip in *Memories*, chapter IX, iii, which also gives more detailed accounts of some of the experiences mentioned in this lecture.

[307] Jung's work abounds with references to Lévy-Bruhl's notion of *participation mystique*, or mystical participation: "It denotes a peculiar kind of psychological connection with objects, and consists in the fact that the subject cannot clearly distinguish himself from the object but is bound to it by a direct relationship which amounts to partial identity" (*Types*, Definitions, no. 40).

robbed him, for example: Someone seizes a weaker man's planting patch, who thereafter threatens to take his life and haunt the perpetrator in the guise of a spirit. Often, the thief subsequently returns the land. Just as often, however, avarice proves to be stronger, so that the victim of the theft climbs up a tall tree and leaps to his death.

These people can readily identify with one another and transfer something onto someone else. Thus the field of human objects is missing in a way, because it is already contained in one's own I-plurality. It is as if you observed a school of black fish in the water: when one of them suddenly changes direction, all the others do so as well. Therefore, they possess a veritable group consciousness that reacts as if the whole group had been affected. That is why they are easily given to panics like the "stampedes" of wild herds. They possess a pronounced "mob psychology."[308]

The individual is only the whole. They do not reflect upon themselves, thus making it difficult for us to get along with them. When one primitive adopted a position like Rodin's "Thinker,"[309] he was asked, "What are you thinking about?" He jumped up furiously, and exclaimed, "But I am not thinking at all!" Primitives do have dreams, but are not really conscious of them.[310] Just as if I were to ask you here whether you have any religious customs? "Of course not!" you might exclaim. But then I happen to visit you at Easter and you are just hiding eggs in the grass. "Well, it's just what one does!" One scholar once observed, "This goes to show how primitive these [indigenous] people are. They don't even know what they are doing!" But do you know what the meaning of an Easter bunny or a Christmas tree is? None whatsoever! Nothing other than: "Well, it's always been what people do."[311]

But one matter is certain: these people have a classification system. One man stands out: the chief. Members of the tribe stand side by side, but the chief does not belong in this row. He possesses "mana." He stands before the people. A medicine man from a tribe in Central Africa[312] once

---

[308] "Stampedes" and "mob psychology" in English in the original notes.

[309] The famous bronze sculpture by Auguste Rodin (1840–1917).

[310] Jung also mentions this episode in the summer semester of 1934; see Volume 2 (forthcoming).

[311] These are favorite examples of Jung's to show that archetypal images "had simply been accepted without question and without reflection, much as everyone decorates Christmas trees or hides Easter eggs without ever knowing what these customs mean. The fact is that archetypal images are so packed with meaning that people never think of asking what they really do mean" (Jung, 1934b, § 22).

[312] So Sidler; according to the notes of M.-J. Schmid, it is the chief himself, not the medicine man, who explains these matters.

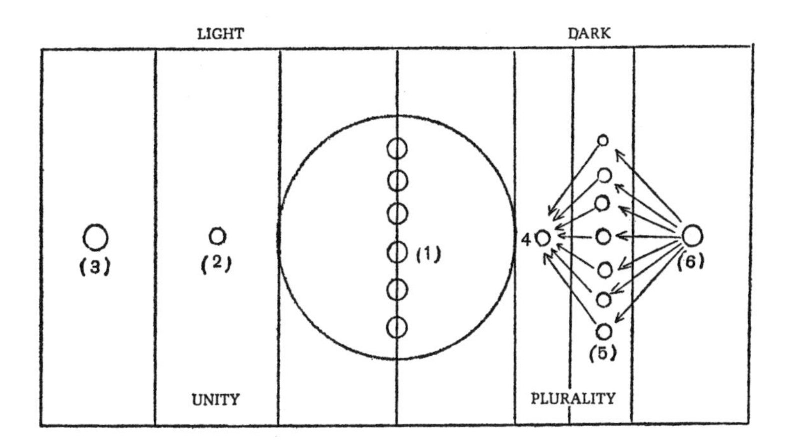

told me in a palaver about the position of the chief compared to the other members of the tribe. In doing so, he used a bundle of small sticks, which always accompany a proper palaver—for each affair discussed, a small stick is stuck in the ground. "The chief is like this . . ." he said and placed one small stick in the ground, "and the others are like this . . ."—and he placed the remaining sticks in the ground in a row. "The chief has mana."[313] The chief thus equals the king.

Primitives are unaware of their own "I." There are many people among us, however, who are also unaware of their own I. Many neurotics have no consciousness of their I at all and are completely identified with their environment. "Oh, what would my parents say?" Or their aunt or anyone else for that matter. Their only standard is what others think, but it never occurs to them what they themselves might think about it. A young lawyer once said to me, "You know, one can do anything provided other people don't know it." Now, this fellow was unaware of his own moral motive, because he had no knowledge of it whatsoever.

When we come back to the dark side of the diagram, there is again an empty circle because the I is missing, but further back a single figure stands out again: the medicine man, who has immense influence as the interpreter of the spirit world from which he draws his power. Actually, this dark sphere has no definite and separate existence, because we are at a loss to say if the right side has been taken for the real side, or the other way around. Reality is probably rather in the background, because

[313] Sidler noted here: "Also see the chapter 'Archaic Man' in Dr. Jung's book, *Seelenprobleme der Gegenwart*, Verlag Rascher, Zürich" [1931b]].

the latter is much more differentiated. With primitives, it is never clear whether this is a dream or reality. For example, a native had a dreadful nightmare. He dreamt that he was captured by his adversaries and burned alive. The next morning, he demanded that his relatives burn him alive. They refused at first; ultimately they agreed to tie him up, and to lower him into the fire by his feet. Thereafter, he hobbled around for nine months until the burns had healed—and all for fear of the dream.[314]

There is a multitude of ghosts, who are all connected with the dark principle. This dark principle is *ayík*.[315] When I tried to speak to the natives of this dark God as of a second God, they protested: "No, there is only one God!" Then I realized that only one reigned at a time: From six o'clock in the morning to six o'clock in the evening, *adhísta*, the benign God, the good and bright spirit, rules; and from six o'clock in the evening to six o'clock in the morning rules *ayík*, the uncanny, dark, evil God. What is true for the day is reversed at night. Beauty abides by day while terror prevails at night. This is due to the fact that when you bask in the tropical sunshine, after a while you will feel as though you were drunk. When the outdoor temperature exceeds your body temperature, you will find it increasingly difficult to imagine anything unpleasant that could really agitate you. You become indifferent to everything. I experienced this myself in Africa as I lay in my hammock, hardly finding the energy to light a pipe. I thought hard of all my most depressing problems to see if they would affect me, but I remained absolutely indifferent.[316] No sooner has the sun gone down, however, than this optimism of the day immediately keels over into the absolute pessimism of the night.

---

[314] Jung had read this story in Lévy-Bruhl (1910, p. 54), who himself had quoted the Jesuit missionary Père Laul Lejeune and his *Relations de la Nouvelle France*. Jung told this story also elsewhere (1928, § 94), but in both instances embellished it somewhat—there is no mentioning in Lejeune of relatives, or that they at first refused to burn the dreamer, and it took him six months, not nine, to recover: *Un [primitive], ne croyant pas que ce fût assez déférer à son songe que de se faire brûler en effigie, voulut qu'on lui appliquât réellement le feu aux jambes de la même façon qu'on fait aux captifs, quand on commence leur dernier supplice. . . . Il lui fallut six mois pour se guérir des ses brûlures* [A primitive who believed that in order to comply with his dream it would not be enough to burn himself in effigy, requested that his legs be actually burned in fire, in the same way as one started to burn prisoners to death. . . . It took him six months to recover from his burns].

[315] Sidler notes: "the word expresses a sudden cold gust of wind."

[316] Jung recounted this also in the German Seminar of 1931: "The indigenous people lie around in the sun as if they were drunk. I observed this intoxicating effect of the tropical sun in myself: one is no longer touched by anything unpleasant, and is overcome by boundless indolence. (For example, I could observe, with the watch in my hand, that it took me a whole hour to reach the decision to light my filled pipe.) The climate produces an immense optimism" (forthcoming).

Lecture 12

26 JANUARY 1934

I WOULD LIKE to draw your attention to an excellent opportunity to admire the works of a recently deceased artist, Otto Meyer,[317] including some small pictures of mandalas, at the *Kunsthaus* in Zurich. We can assume, I suppose, that he was not familiar with such matters, but rather that these pictures arose from his own visions. Some of these works are reproduced in the exhibition catalogue. You will find typically octagonal shapes featuring a human figure at the center.[318] Octagonal light is a symbol of the unconscious symbolic source of light. One therefore speaks of il-lumination,[319] a frequent occurrence in mysticism. The figure is either the visionary himself, or a figure that emerges from this light. We shall return to this later.

Today, I would rather turn to another case, in asking you to bear in mind the schema introduced last time. This diagrammatic presentation never appears as a whole in an individual case; rather, it is as if a light beam were moving along a dark elongated scale. With Mrs. Hauffe, it was the extreme left end of the scale that stood in bright light. In the present case, the light beam moves to the right, whereby the psychological image changes considerably.

This case concerns a person who played a not inconsiderable role at the turn of the century, a lady who lived in Geneva under the pseudonym "Hélène Smith." Professor Théodore Flournoy published a book about her with the peculiar title, *From India to the Planet Mars: A Case*

[317] This refers to the memorial exhibition for the Swiss painter and graphic artist Otto Meyer-Amden (1885–1933), from 22 December 1933 to 28 January 1934. The exhibition was also shown in Basel and Berne. The catalogue: *Gedächtnisausstellung* etc., 1933.

[318] See for instance catalogue numbers 252, 254, and 257. Sidler went to the exhibition two days later, made sketches of those pictures, and added them to his notes.

[319] Original: *er-leuchtet*; *lumen* = Latin for "light."

of Multiple Personality with Imaginary Languages (Geneva, 1900).[320] The main title would appear to have been his publisher's, the subtitle his own.

"Hélène Smith" was born to a Hungarian father, while her mother was presumably a local Genevan. The father was highly intelligent and well-educated, an excellent linguist, and spoke fluent German, French, Hungarian, and English; he also had some knowledge of Greek and Latin.

Concerning the peculiar phenomena that occurred to her, Hélène appears to have taken after her mother. On account of various strange experiences she had had herself, her mother had become a convinced spiritualist. When one of her daughters was three years old, for example, the mother saw a white figure standing by the child's bedside. The child was taken ill the next day and died. Hélène's brother also related peculiar events. He would hear footsteps and experienced other strange phenomena. It seems that the family was rather poor. So much for the characteristics of the milieu that had a strong influence on Hélène's character.

While her exact date of birth is unknown,[321] she was baptized a Catholic, but later raised as a Protestant. She had an excellent character, was intelligent and a good pupil, although she was sometimes inattentive in school, because various fantasies interfered with her concentration. When she was fifteen years old, she began an apprenticeship at a *Grand Magasin* [department store], where they thought very highly of her. She was occupied with her daydreams, among which she had all kinds of favorite ones, which nearly bordered on visions. No neurotic symptoms proper seem to

---

[320] The book documents the experiences and séances of "Hélène Smith" (real name Catherine-Elise Müller [1861–1929]), a famous late-nineteenth century psychic. It was published at the end of 1899 (at Christmas; see Flournoy, 1914, p. 75), but postdated to 1900, exactly like Freud's *Interpretation of Dreams*. While it took the latter eight years to sell the first print run of 600 copies, however (Jones, 1953, p. 360), Flournoy's book caused a sensation and after only three months went into its third edition (Shamdasani; in *Planet Mars*, p. xxvii). Théodore Flournoy (1854–1920), professor of psychology at the University of Geneva, made a great impression on Jung, who also offered to translate the book into German, but Flournoy told him that he had already appointed another translator. In him, Jung found "above all someone with whom I could talk openly. . . . [H]e soon represented to me a kind of counterpoise to Freud. With him I could really discuss all the problems that scientifically occupied me" (ibid., p. ix). Jung made a tribute to him that was appended to the German edition of *Memories*, but not included in the English one. The first English translation appeared in the 1994 edition of *Planet Mars*.

[321] That is, to Jung. All the information that Jung gives on "Hélène's" childhood and youth, and her initiation into spiritism, are taken from chapters 2 and 3 of Flournoy's book (which does not give her date of birth). I have chosen not to encumber the footnotes with page references to each and every piece of information; they can be easily found by following the narrative of the book.

have occurred, however. Already as a child, she depicted her dreams in embroideries, drawings, and paintings, showing great talent. She engaged in these visions actually as a form of entertainment, as a pleasant game. She derived particular pleasure from drawing landscapes, buildings, and paintings, all in bright, rosy colors.

When she was fourteen or fifteen, that is, during puberty, she had highly impressive visions. In one of these, she saw a bright light, and in this light strange letters whose meaning escaped her. On several occasions, a somewhat scary man appeared next to her bed, dressed in a strange, colorful robe. She also frequently reported that a man had followed her on her way home, but the objectivity of these accounts could not be verified. Then it happened to her that certain disturbances suddenly appeared in her handwriting: letters were replaced by strange signs. She felt increasingly surrounded by a strange protective spirit. When she was ten years old, she was attacked by a dog, whereupon a monk appeared who chased the dog away. She described him as a man clothed in a long, brown habit, with a white cross on his breast. Later, too, when a man made advances, for instance, the figure of this monk would reappear. When she was about twelve years old, she used to start whenever the doorbell rang, because she believed that the great event that she expected had indeed occurred—namely, the arrival of an elegant carriage drawn by four white horses, from which a noble gentleman wearing a gold and silver embroidered coat would alight to collect her and lead her to the faraway land where she really belonged. This is the familiar infantile fantasy of noble descent, that is, that the child thinks she has only surrogate parents.

During those years, she had a pronounced fear of the world and the "outside" in general. She shunned people and avoided social occasions, concerts, parties, and so forth. She suffered constantly from a vague discontent and was never truly happy. Her pride and her ambitions chafed against her humble surroundings. Her taste belonged to a higher sphere; she was too strong a personality to fit in her milieu.

Once, when she was feeling unwell and had to consult a doctor, he made advances and attempted to kiss her. At this moment, the monk reappeared and "effectively interrupted" the situation, as she recalled. She had been stricken by a dreadful fear of everything at the time. Towards her twentieth year, her condition changed for the better. The fear abated, and she was successful at work. She was business-savvy and dependable, a valued colleague, and it seemed as if she could go in for a career. One might also have expected her to get married sooner or later. But whenever a situation occurred that might have aroused pleasant feelings, this unpleasant protective spirit reappeared and whispered: "No, this isn't the right one yet." The right one would only come when the time was ripe. Thus, he interfered with all such situations and actually made it impossible for her to experience life in a natural way. In this way, she began to prepare herself for a career as an old maid. Hélène was quite vivacious, however, and the situation had to come to a head. Somehow her temperament had to break through. This also accounted for her dissatisfaction and her conduct, which, however, her surroundings failed to understand.

Through her mother, whose interest in spiritualism I mentioned above, Hélène joined a spiritualist circle in 1892, the group of Mme. N. The circle met in Geneva, and consisted of people from all walks of life, some well educated, others not, and the whole spectrum in between. Hélène soon established herself as the group's leading medium. It quickly became apparent that she was much more strongly directed by the unconscious than the others. As we know, the messages conveyed by spirits are usually awfully trivial. Accordingly, Hélène reported nothing special. At first, the spirits conveyed their messages through automatic writing. But she soon entered a trancelike, somnambulistic state in which she had visions. Shortly before, she had seen a balloon go up, and now a balloon appeared to her, which was sometimes bright and sometimes dark. Then the balloon disappeared, and bright ribbons appeared instead, from which a star emerged, whose bright radiance filled her field of vision. Here we have a vision of the inner source of light, of the symbolic light. Thereafter, this star turned into an awful grimacing head with flaming red hair, a devil of sorts. The figure's red hair turned into a bouquet of red roses, out of which a small serpent wriggled. These were her first visions in that spiritualistic circle.

Such visions are highly typical. The things that she sees obviously correspond to the spectrum of inner possibilities: all kinds of fantasies, the spirits, and finally the light. At first she sees this brightly lit balloon, and naturally we would say: Well, she did watch a balloon go up shortly before the vision, and this impressive spectacle was reproduced here. In reality, however, there is no such causality here. It is merely the same word. What she really sees is the light, and she thinks of the balloon because she has just seen one, and this is the nearest analogue to what was seeking expression in her consciousness. If we dream of a locomotive, then we do so because it refers to something that could perhaps be compared to a locomotive.

Whenever a positive vision occurs, we can expect matters to immediately turn around and turn into their opposite, in accordance with a psychological law, Heraclitus's principle of "enantiodromia."[322] It denotes the reversal of events, that is, their turning into their opposite. So when a very bright vision occurs, it will soon be followed by something dark, something good by something bad, right by left. So the brightly lit balloon gives way to a dark one, the radiant star to a hideous devil's face. The star has an elevating quality, it is a beautiful sight, which arouses positive feelings; but in accordance with the principle of "running counter" it turns into something hideous, despicable, and evil. In turn, these qualities shift to the positive, becoming the bouquet of roses. Then likewise again, this becomes a dangerous serpent. A serpent lying concealed beneath flowers is a familiar poetic image.

Thereafter, Hélène also had visions at home. The man who haunted her regularly now appeared in a white coat and a turban. Other strange events occurred, which could also be verified. At her workplace, for instance, a certain pattern was missing. Although the clerk looked for it everywhere, he was unable to find it. Observing his quandary, Hélène remarked: "But it was given to Mr. J., wasn't it?" The clerk looked at her, laughed, and said: "But you couldn't possibly know that!" At that moment, the number "18" appeared before her, and she said: "Eighteen days ago, this pattern was lent to Mr. J.; please check!" And indeed, upon checking the books, what Hélène had said proved true. People suggested that this was just a [random] idea that crossed her mind, because she could have had no actual knowledge of the matter. But you never know!

In the séances, she gradually developed a high degree of somnambulism. The guide and shadow spirit came into action. About this time she also became acquainted with Flournoy, who was interested in such phenomena. Already at their first meetings he observed a one-sided anesthesia of the right side of the body at the onset of somnambulism, whereas the left side became hypersensitive. When pain was inflicted on her right hand, she did not feel any pain; after a while, however, she felt the pain in the corresponding spot of her left hand.

---

[322] Greek, to run counter. This is a key concept in Jung's thinking. Throughout the numerous references to it in his works, Jung attributes the term to Heraclitus of Ephesus (ca. 550–480 BCE). Although the concept is in accordance with the latter's philosophy, it seems that the term itself was not used by Heraclitus himself, but first turned up in a later summary of his philosophy by Diogenes Laërtius.

Her protective spirit or "control"[323] could be induced to speak through her if one addressed the right-hand side of her body; she would respond by tapping her left hand. While she consciously conversed with other people, she could indirectly stay in touch with this unconscious figure that delivered complicated messages. It was also observed that in a hypnotic state she showed the phenomenon of "allochiria," that is, a confusion between left and right. In this state, she steadfastly insisted that Rousseau Island is situated on the left side in the Rhone River on the way down from Geneva station, whereas in actual fact the island is on the right. When someone lifted her left arm, she said it was her right one.

These happenings illustrate the psychic process. In terms of the diagram, the light beam has shifted to the left side, and the right one has become anaesthetized. The personality shifts to the unconscious side. Through this leftward shift the entire image folds onto the axis, thereby activating the protective spirit or control. The control was a male figure, and behaved "very personally." During the sessions, he usually took hold of her right hand and wrote with it, while speaking to her from the left-hand side. When he wrote, his handwriting was completely different from the medium's. He also had a male voice although he spoke through the mouth of the medium. He behaved entirely as his own master, and acted completely independently of the wishes of the medium and the séance participants. His character was at first rather vague, but he soon developed distinct characteristics. For instance, he had a flair for poetry and magniloquently called himself Victor Hugo.[324] He wrote a prolific amount of verses, but, alas, they were not by a Victor Hugo! His verse was overly sentimental and trivial. After he had ruled for five months, his antagonist entered the scene, a man with an Italian accent and a coarse voice. He called himself Léopold. He was rude, uncouth, and interrupted séances. Now Victor Hugo got all worked up about him. Léopold disdained the whole group, and would have liked to blow up the whole thing, and in particular he was intent on ousting Victor Hugo whose verse he ridiculed. Léopold was jealous, vindictive, and evidently besotted with Hélène whom he courted shamelessly. He had a pronounced dislike against the mixed-gender composition of the group, which, it must be said, took a fair share of liberties with the medium. His manner was arrogant and overbearing, and he was determined to be sole master of Hélène whom he did not want to share with anybody.

[323] This word in English in the notes.

[324] Victor Hugo (1802–1885), the famous French poet, novelist, and dramatist.

Flournoy believed that Léopold had arisen from some kind of "autosuggestion."[325] We have to grant him that this already happened in the year 1899, at the heyday of autosuggestion. At the time, "autosuggestion" was still an excellent term, and was welcomed as an explanation for all kinds of things. Such terms, however, quickly waylay thinking. Everyone believes, "Ah, now something has happened!" But nothing whatsoever has happened, apart from someone mentioning the term "autosuggestion." So one assumed that the medium had unconsciously taken it into her head to invent a ghost named Léopold.

In the 1840s, a scholar visited a tribe of North American Indians. He spoke to an old chief, who told him about the peculiar experience of having visions. Being an enlightened man, our scholar had already heard about these "imaginations," and said to the chief: "It is not really like that; you just imagined it!" Whereupon the Indian replied: "Well, but *who* imagined it in me? Because that's how it is, isn't it, someone must have imagined it for me!" We, too, say, "I imagined something." But we are just cautious in saying "I." It would be too frightening if something existed in our soul that is autonomous, something that imagines itself in us; in other words, that there is something in our own inner sphere that can act regardless of our wishes. This is an uncanny thought, just as uncanny as believing that someone were under our bed.

Léopold had always existed. He already existed in the figure of the monk. He was part of Hélène's psychic structure. It is merely a question of the distribution of light. Such or similar figures become active in all of us; they are very typical figures.[326] William James, the famous philosopher and pragmatist and the elder friend of Flournoy, understood this fact much better. He maintained that "thought tends to personal form."[327] All psychic content tends to become personal under certain circumstances, on condition that the light move towards the left, so that consciousness moves slightly from right to left. Immediately the thoughts will take a personal form and become autonomous. This is precisely what happens when a medium falls into trance. Everything becomes reversed, because the inner image surfaces and the control is in command. Mediums will then receive all kinds of information that they could not possibly have known before, often of a sinister character. The Latin word "sinister," after all, denotes both "left" and "uncanny."

---

[325] See, e.g., *Planet Mars*, p. 59.

[326] M.-J. Schmid notes in parentheses: "Animus."

[327] "Thought tends to Personal Form" (James, 1890a, p. 225). Also quoted by Flournoy in connection with autosuggestion (*Planet Mars*, p. 59).—On James see note 179.

Lecture 13

2 February 1934

Submitted Questions

I have received a few reactions, probably from some of the younger members of the audience,[328] wishing me to present fewer case histories, and instead give you more of my own point of view. Now I consider that I have done this pretty freely already, but you must bear in mind that I set out to give a course of lectures on modern psychology, and I cannot claim that modern psychology is identical with myself. It would be most immodest if I advanced my own views and opinions more than I already have. On the other hand, such factual material is an indispensible component of such lectures; we have to deal with the whole of the human soul, and we must remain close to everyday life so as not to get lost in cold theories. So if we discussed solely the problem of the transcendent function,[329] for instance, we would all be parched from too much spirit. Compare the study of anatomy: presumably, we must take into consideration the entire human body rather than limit ourselves to one opinion about it.

I should like to repeat that the cases that we have been discussing are not unique or abnormal, as some of my audience still seem to think, but typical. If you dissect a salmon in a laboratory, you are not studying one salmon in particular, but simply *the* salmon. So these experiences lie more or less hidden in the unconscious of ordinary people. But it is true that very few are able to actually have these experiences, and so it is an exceedingly difficult task to make them comprehensible. My psychology

[328] See **note 162**.

[329] A concept Jung had introduced in 1916 (1957 [1916], §§ 166 sqq.), denoting "the collaboration of conscious and unconscious data" (ibid., § 167), later termed "active imagination" (e.g., 1968 [1935], §§ 390 sqq.). It describes the emergence of a "sequence of fantasies produced by deliberate concentration" (1936/1937, § 101). This method emerged out of his self-experimentation, in which "he deliberately gave free rein to his fantasy thinking and carefully noted what ensued" (Shamdasani, in Swann, 2011, p. xi), entering into a dialogue with the visionary figures he encountered.

comprises, after all, quite a number of concepts that are underpinned by experiences that are not generally accessible. If one posits a very general claim, which has been verified in many instances, one can always hear people exclaim, "Well, quite the contrary is true in my case!" If you seek to explain to such people why their case is different, you will need half their life story. Consequently, I must set out some fairly broad factual material. This is a necessary prerequisite for the study of modern psychology, for these are cases that render visible what is invisible in your specific case. Such things do occur, however, and we must therefore take them into account.

* * *

Let us now return to our case and to the figure that I started discussing in the last lecture. We encountered a male figure named "Léopold," a kind of protective spirit, an animated shadow, that acts of his own accord, and from time to time intervenes in critical situations, which is referred to as a "teleological automatism."[330] This is a scientific term, however, that is somewhat too general. This helpful or disruptive intervention shows intention and intelligence, and thus cannot be considered an automatism.

This figure had not existed before Flournoy began observing Hélène. Victor Hugo existed beforehand, and the figure of Léopold emerged only later. At one of the séances of the Group N., Hélène had a vision of a conjuror showing her a carafe of water and pointing at it with a small magic wand.[331] One of the members of the circle interpreted this figure as "Balsamo"—which is another name for "Cagliostro," since this very *scène de la carafe* occurs in a novel by Alexandre Dumas père, *Mémoires d'un Médecin, Josef Balsamo*.[332] The protective spirit himself added that "Léopold" would only be his pseudonym, and that in reality he would be Balsamo or Cagliostro, under which name this famous eighteenth-century magician and impostor had fooled the world. This episode then gave rise to an entire novel. Hélène imagined that she had been Balsamo's medium, Lorenza Feliciani, in a prior existence, and would now be her reincarnation. It turned out, however, that such a figure had never existed in reality, but had been invented by Dumas. Léopold then explained that

[330] *Planet Mars*, pp. 22, 41 sqq., 57, 83. On the figure of Léopold in general, see ibid., chapter 4, "The Personality of Léopold," pp. 51–86.

[331] Ibid., pp. 60–61.

[332] Dumas, 1846–1848 [1860]; the scene with the carafe of water in chapter 15. Dumas was inspired to write this novel by the life and personality of the Count Alessandro di Cagliostro (1743–1795), an alias for the occultist and adventurer Giuseppe (or Joseph) Balsamo.

she would have actually been Marie Antoinette, which again refers to the *scène de la carafe*.[333] This episode in Dumas's novel describes an accidental meeting between Balsamo and Marie Antoinette of Austria at the castle of Taverney, where she stayed on her way to Paris. She had been told that he could tell the future from a glass of water. Scrying is used all over the world, by the way, also by primitives, to inspire intuitions. So Balsamo looks into the carafe and sees her destiny, but refuses to tell Marie Antoinette what he had seen. Upon her steadfast insistence, he eventually yields and tells her to kneel before the glass and look into the carafe after he had waved his wand over it. She has a dreadful vision: she sees her own execution and faints. Hélène believes that she is a reincarnation of Marie Antoinette, whereupon a love affair ensues between Balsamo / Léopold and Marie Antoinette / Hélène; however, this is not borne out by historical evidence.

Flournoy was preoccupied with the name of Léopold, and justly so, because such names often have a deeper meaning. A colleague drew his attention to the fact that "Léopold" contains three consonants, LPD, and that the same three letters represented the initials of the motto of the Illuminati, a secret society: *lilia pedibus destrue*, "destroy the lilies with your feet." In the beginning of his novel, Dumas describes a gathering of Illuminati and Freemasons from a great many countries on Mount Donnersberg near the German town of Mainz in May 1770, presided by the famous visionary Emanuel Swedenborg.[334] A stranger appears and asks to be admitted to the society. He then reveals himself to them as *celui qui est*, "The One Who Is." The president confirms that the stranger is the "Illuminated One," since he bears the three letters—"L.P.D."—on his chest. The envoys of all countries recognize him as their leader and await his instructions. With their assistance, he intends to overthrow the monarchy within twenty years, and to establish a new world order. To this end all envoys must follow the motto *lilia pedibus destrue* in their home countries, that is, destroy the lily, the emblem of the Bourbon monarchy.

---

[333] Marie Antoinette, born Maria Antonia (1755–1793), daughter of the Austrian Empress Maria Theresia, archduchess of Austria, later dauphine of France (1770–1774), and, as wife of Louis XVI, queen of France and of Navarre. In the French Revolution, she and her husband were executed by guillotine in 1793.

[334] Emanuel Swedenborg (1688–1772), Swedish scientist, philosopher, revelator, and mystic; best known for his book on the afterlife, *Heaven and Hell* (1758). There are repeated references to Swedenborg in Jung's works, e.g., to his vision of a fire in Stockholm, which allegedly proved to correspond to the facts (Jung, 1905b, §§ 706–707; 1952, § 902). Swedenborg's experiences instigated Kant to anonymously write his *Dreams of a Spirit-Seer* (1766), quoted by Jung earlier (see p. 42).

The Illuminated One himself travels to France, where he begins to prepare the Revolution.[335]

Historically speaking, this is not quite accurate. The gathering is antedated, since the order of the Illuminati was not in actual fact founded until 1 May 1776—namely, by Adam Weishaupt, a former Jesuit who later became a Freemason.[336] The society sought to pursue the ideas of the Enlightenment, and to bring about liberalism and free-thinking, bringing it in conflict with the prevailing social order, and the Illuminati were subject to frequent persecution. One of Weishaupt's first collaborators, by the way, was Baron Knigge, the author of the famous book, *On Human Relations*.[337] Towards the end of the eighteenth century, the movement abated until its revival in Germany in 1880.

These literary parallels could leave us cold, if they were not psychologically meaningful, that is, symbolic. Seen in this context, it becomes clear that this spiritual leader or mentor, "Léopold," actually is a member of a secret order, that is, an association of unknown leading thinkers. It is significant that they were Illuminati, because their member lists contained all sorts of famous names, like Herder[338] and Goethe. This is an important psychological criterion, for Léopold represents not a unity but a multiplicity. He has replaced neither Victor Hugo nor Balsamo, but represents all of them: he is the poet, the magician, and the member of a secret order. When he was asked himself about the origin of his name, he replied that he had taken it from a friend who belonged to the House of Austria.[339]

---

[335] Cf. Dumas, 1860 [1846–1848], chapters 2 and 3, particularly pp. 15, 18, 31.

[336] Johann Adam Weishaupt (1748–1830), German philosopher and jurist. The society of the Illuminati was banned by the government of Bavaria in 1784; Weishaupt lost his position at the University of Ingolstadt and fled Bavaria.

[337] Adolph Freiherr Knigge (1752–1796), German writer and Freemason. In 1780, Knigge joined the Bavarian Illuminati, but withdrew in 1784, due to dissensions with Weishaupt. His book (1788)—actually a treatise on the fundamental principles of human relations—enjoyed immense success, and in German the word "Knigge" still stands for "good manners" or (books on) etiquette.

[338] Johann Gottfried (after 1802: von) Herder (1744–1803), important German philosopher, theologian, poet, and literary critic. He met the young Goethe in 1770, and inspired him to develop his own style. Later, Goethe secured him a position in Weimar as General Superintendent.

[339] While at first "he immediately accepted . . . the hypothesis of M. Cuendet" (*Planet Mars*, p. 291) about the significance of LPD, he seemed to have forgotten this explanation when he was questioned again later, and explained "that he took as a pseudonym the first name of one of his friends from the last century, who was very dear to him, and who was part of the house of Austria though he didn't play any historical role" (ibid.).

In the course of subsequent meetings, he acquired a particular technique: he would take hold of the medium's hand, and while séance participants were speaking to her, he would write with her hand. Often a long struggle ensued over the control of the hand, because Léopold would hold the pencil between the thumb and the index finger, whereas she always held it between the middle and the index finger.[340] Eventually Léopold would gain the upper hand. His writing also differed from the medium's: he used obsolete eighteenth-century handwriting and spelling, although Cagliostro's original handwriting bears no resemblance to Léopold's. The same happened to the voice: although Léopold could neither speak nor understand a word of Italian, he spoke with a coarse Italian accent. Hélène's own voice was not deep, but he taught her how to speak in a deep voice.

Interestingly, he always gave evasive answers to precise questions, whereas he flowed over with moral and philosophical talk, and wrote verse in the manner of a "Victor Hugo *inférieur*." His memory was better than Hélène's, in particular for certain data and other matters that usually easily escape one. Repeatedly, Hélène felt to be actually identical with him. Although he behaved differently, she often felt as if this figure had entered her being and overwhelmed it, resulting in a loss of her identity, in particular at night and in the early hours of the morning. The two states of consciousness were not completely separate, and Hélène and Léopold shared certain peculiarities, for instance animosity. Hélène was rather nervous and easily excited. Cagliostro/Léopold had quite a temper, was choleric, sometimes most unpleasantly brusque, and irritable, and also overtly enamored of himself. He was steeped in forthright animosity, and claimed to be an authority in all possible areas—and in those areas where he was not, he overrode his uncertainty with even more assertive statements. In particular, in the guise of Balsamo he prided himself on being a physician and alchemist who knew about elixirs and secret remedies. Many people consulted him through Hélène, although he affected a great disdain for modern medicine, and his prescriptions were as antiquated as his spelling. It then turned out that Hélène's mother was well versed in the curative powers of herbs and plants.

Léopold's tender love for Marie Antoinette, that is, Hélène, played a great role. He wrote her affectionate letters and poems. Curiously, he also felt affectionate towards Flournoy, referring to him as *mon ami*, my

[340] Ibid., p. 100.

friend. He and Hélène were Siamese twins in a sense, since she, too, had tender feelings for her doctor, although she was not aware of this. On the other hand, Léopold was dreadfully jealous, and made Hélène the most amazing scenes when a male member of the group paid her any attention. Flournoy says of him: *ce mentor austère et rigide, . . . présente, en somme, une donnée psychologique très générale; il n'y a aucune âme féminine bien née qui ne le porte logé dans un de ses recoins.*[341] Her letters also attest to her peculiar affinity with this figure: all of a sudden, a word or a whole sentence would appear in his handwriting, and in antiquated French. And he also appeared in her dreams.

For the moment, I shall not continue with my description of this figure's character. Pursuant to your requests, I shall instead set forth my own views. This figure is by no means unique, on the contrary it is very quite common, only we do not often see it in such a definite form. Flournoy's description is classical. Although he had no idea of what he was describing, he intuited something, and gave a most clear description of the case. It is a typical and universal figure which I have called the "animus," and no woman exists who does not possess it. I know of only one case where this figure was absent, and this was a militant suffragette, a friend of Mrs. Pankhurst's.[342] She did not have this figure because she was it herself! The only other case that I recall was a hermaphrodite who came to me because she was in doubt as to whether she should live her life as a man or a woman. However, the number of those is legion who lack any knowledge of this figure. It is very difficult to point it out in many women because it is not clearly dissociated, but remains utterly in the dark and so naturally manifests itself only indirectly. One must turn to the shadow consciousness[343] to prove the existence of this figure. To do so, I must draw another diagram on the board. I am not sure whether it will meet with your approval, but so be it.

---

[341] "This austere and rigorous mentor . . . represents, in fact, a very common psychological attribute; there is no noble feminine soul that would not carry such a mentor in one of its recesses" (*Planet Mars*, p. 82; trans. modified).

[342] Emmeline Pankhurst (1858–1928), British political activist and leader of the British suffragette movement. Although widely criticized for her militant methods (smashing windows, assaulting police officers, arson, etc.), her work is recognized as a crucial element in achieving women's suffrage in Britain.

[343] "Shadow consciousness" [*Schattenbewusstsein*] is not a technical term, and does not appear in Jung's other writings. The concept of the "shadow" as such, however, of course plays a prominent role in his theory. He later defined the shadow as the "'negative' part of the personality, i.e., the sum of the hidden and unfavorable qualities, of the insufficiently developed functions, and the contents of the personal unconscious" (Jung, 1917–1942, § 103[5]; note added in 1942; trans. mod.).

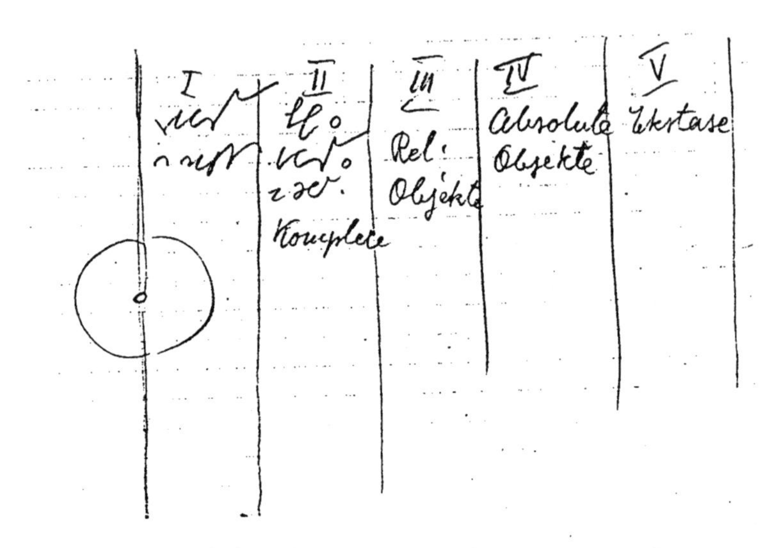

I: Unconsciousness of the subject; II: Stage of consciousness of the contents; complexes; III: Religious objects; IV: Absolute Objects; V: Ecstasy

We can differentiate between five spheres or stages. In the first sphere the "shadow consciousness" begins to make itself felt. In most persons it manifests itself as a slight feeling of missing something, or that something is not quite as it should be, as a *léger sentiment d'incomplétude*,[344] which gives rise to self-consciousness and a sense of inferiority. People look for the cause of this disturbing feeling in the outer world; they think perhaps their collar is crumpled or their tie crooked. Or a scholar has inferiority feelings because he feels his latest book is not quite as good as he thought, but will be terribly touchy should someone else criticize it. A tenor, of course, will have symptoms in the larynx, and an officer of the infantry in the feet. They place the inferiority where they really need not fear criticism; but in analysis I have to show them that their real inferiority lies somewhere else entirely. What we lack is never what we think it is, just as neurotic inferiority feelings never spring from where we claim they come from, but from real inferiority.

[344] Roughly, a feeling of uncertainty or incompleteness—an expression coined by Pierre Janet, to which Jung repeatedly referred, both in his works and his letters. With it, Janet described, in the case of psychasthenic patients, *le fait essentiel dont tous les sujets se plaignent, le caractère inachevé, insuffisant, incomplet qu'ils attribuent à tous leurs phénomènes psychologiques* [the essential fact of which all subjects complain, the unachieved, insufficient, incomplete character they attribute to all their psychological phenomena] (Janet, 1903, I, p. 264). See also note 177 **[on Janet]**.

If the light of consciousness falls onto the right side, only a strange unawareness of subjective motives exists, a feeling of a certain darkness and a faint *sentiment d'incomplétude*. These people believe that all is well in their lives, and they have implicit faith in their good intentions: "All I ever wanted was the best." They come to me with a glowing description of their ideal marriage and happy circumstances; yet I know that a neurosis has brought them, and why should they be neurotic if the conditions of their life are so perfect? These are the people who have no knowledge that this is the place where it is dark.

In the analytical sessions, I slowly move the light of consciousness to the left side, as I endeavor to make each successive sphere conscious. In the first hour, for instance, the light of consciousness moves only into a tiny fraction of the next stage, in the following hour a bit further, and so on.

Even though the shadow makes itself already felt in stage I, the focus of attention is still on the I. When the field of consciousness is limited, the body plays a great role. People who are enamored by themselves are extremely conscious of their bodies, they attach tremendous importance to how they have eaten, slept, digested, and what impression they have made on others. They tend to connect their complexes with the body; inner conflicts appear to them in the form of physical illness. They are egocentric and feel inferior, but dwelling so much on themselves may at least give them some idea of their shadow and their field of consciousness thus tends to become less restricted.

In the second stage the body is still important, but we find not only one object as in stage I, but a number of objects, namely, inner objects. People realize the existence of complexes that are independent of their own I. During the treatment, the light of consciousness shifts further from the right to the left side.

In the third stage something strange happens. People with a very limited range of consciousness, that is, very egocentric people, who have actually never been interested in anything else but themselves, are in fact clearly conscious of their shadow. They can provide you with an excellent description of themselves—if only it were interesting! At this stage, a realization of the contents of the complexes occurs, and they appear as independent, autonomous contents. An altered sense of the body relativizes the objectification; Léopold, for instance, is a relative objectivation. You have seen how this figure overlaps with Miss Smith, and is subjectively tainted by her temperament, character, her view of life, etc.

In the fourth stage, absolute objectivation occurs. In parapsychology, this is to be taken literally. The figures detach themselves and act autonomously, like persons that exist outside of us. These figures have their own will and intentions, and strike us as strange. In the case of the Seeress of Prevorst, for example, the priest is a completely objective entity, and exists in his own right, untainted by the psychology of the medium. Here we encounter those cases where people are preoccupied with figures whom they keep secret, only to suddenly—and inexplicably—commit suicide.

At the fifth stage, the reality of being one's self ceases to exist; it is the stage of absolute reality, of absolute ecstasy. One has changed into something completely different, and the person becomes completely absorbed into a certain absolute existence. This stepping out of reality is precisely what the mystics describe.

Lecture 14

9 February 1934

Today I would like to return to the diagram introduced last time to further elucidate the figures of the unconscious. It consists of ten spheres of consciousness; the five on the right side belong to the consciousness of outer reality, the five of the left side belong to the consciousness of inner reality. Everybody is conscious in some of these spheres.

We saw that in section III on the left side the figure of Léopold shows aspects of Miss Smith, and is colored by the subject of the medium. Now in section IV on the left side we encounter figures that are completely dissociated, and are as autonomous and independent from the subject as another person, a person, that is, who is not subject to my will, but someone who is basically a stranger to me and holds his own views. Thus, something new and strange enters the play here.

At first, we find an intensification of something that had already begun in section III. In the center, in the vicinity of the subject, people are to a large extent identified with their body, and we find feelings that are linked to the body. When complexes begin to interfere and psychological disturbances appear, this link with the body begins to weaken somewhat, and the body is neglected and forgotten due to psychological complications. Here people tend to become more interested in the psychological aspects of their conflicts, and in certain cases they disregard the reality of their body. A patient of mine was so preoccupied with her psychological problems that she once sat down on a bench by the lake to dwell on them, although the thermometer showed minus six degrees Celsius. She sat there for two hours and was surprised that she had to pay for her folly with a severe cold, inflammation of the bladder, etc. Another patient, who had been in analysis for a long time, arrived one day in a completely bewildered state. I asked what the matter with her was. She then got into all kinds of things before I finally had the idea to ask her: "Now tell me, did you actually have lunch?" "No, I forgot!" Hunger and fatigue, particularly when they go unnoticed, avenge themselves by confusing the mind.

As we move through the sections, the body becomes less and less important. In section IV, the reality of the body, its mass, gravity, and undeniable existence, have already become transferred into the object. Here you can see, as in the case of the Seeress, how the figures that preoccupy her have assumed concrete reality. And finally, in section V, the identity with the body and the reality of being one's self cease to exist altogether. Ecstasy means to step out of oneself. Then we are standing outside, are no longer in our body, but have become transformed into something completely different. This is the realm of mystical experiences, whose reality can of course not be doubted, for people have indeed been seized by these things and described their experiences.

In the first sphere there is only one object of consciousness: the I and the body are identical, nothing else exists. In section II there are already several objects, inner objects, that is. These are complexes that have become conscious, objective psychical elements. In the third sphere the complexes become personalized, and the autonomy is even more pronounced. The complex is a very unruly animal, quite independent of our I. Complexes are disobedient, autonomous beings.[345] Everybody who has ever had a complex knows: Something comes to mind although you do not want to think of it, or at night, when you wish to sleep, you are unable to, because the complex is sitting right next to you. Or you want to be particularly friendly to certain people, and yet you fail for the life of you, because the complex keeps putting words in your mouth that you do not intend to utter. Part of absolute reality has already drifted away into the psychic complex, which behaves with almost the same reality as a disobedient dog, or a fly that keeps pestering you—or as Léopold does with Miss Smith.

In the fourth section the objects become even more pronounced, but here they have already become so autonomous and so strange that they can subjugate us entirely. All these inner objects have the tendency to extract the I from its comfortable snail's house. Here, we find those cases of "possession" that are highly common among primitives, and to which I shall return in more detail in a later lecture. Complexes force people

[345] This is a point that Jung stressed over and over again. Although he did not coin the term "complex" itself, he achieved early fame with his theory of autonomous, "feeling-toned complexes of ideas" [*gefühlsbetonte Vorstellungskomplexe*] (1906, § 733; cf. 1913 [1911]), or "emotionally stressed complexes" (Shamdasani, in Jung, 2012 [1912], p. viii), based on his association experiments.

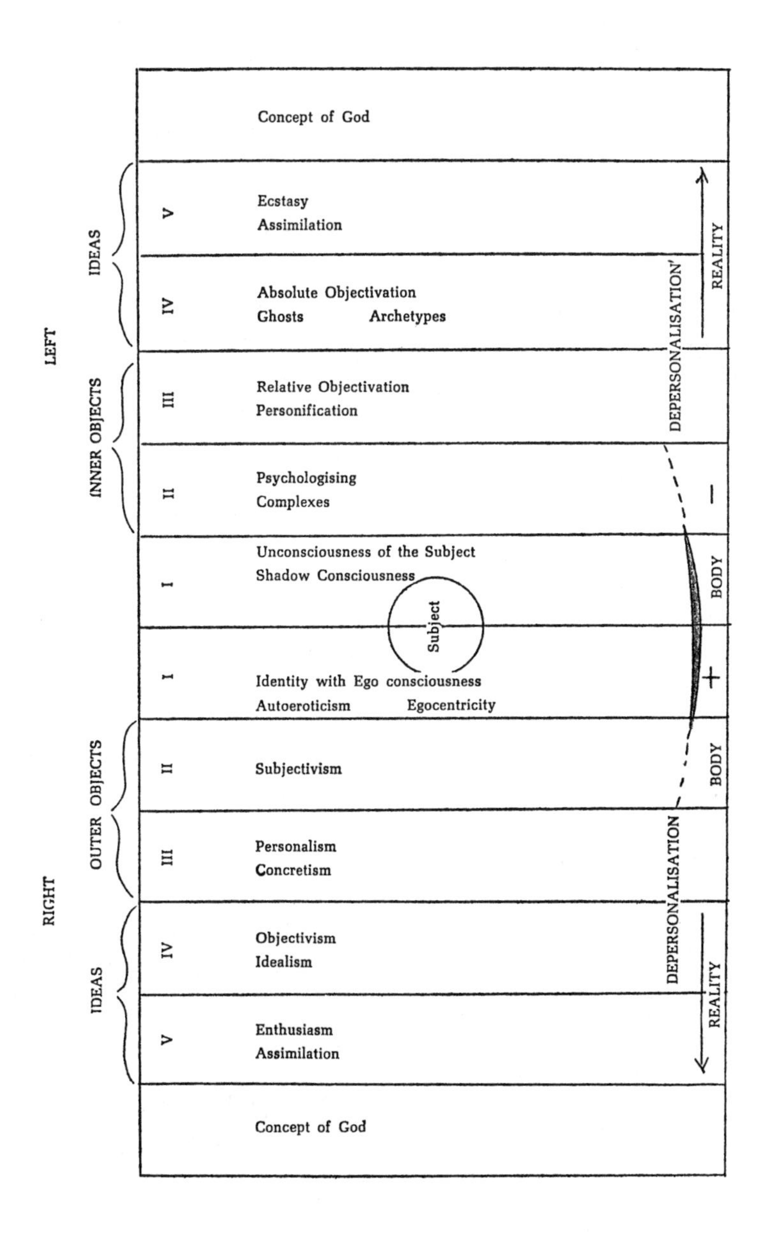
Concept of God
V
Ecstasy
Assimilation
IV
Absolute Objectivation
Ghosts
Archetypes
III
Relative Objectivation
Personification
II
Psychologising
Complexes
I
Unconsciousness of the Subject
Shadow Consciousness
Subject
I
Identity with Ego consciousness
Autoeroticism
Egocentricity
II
Subjectivism
III
Personalism
Concretism
IV
Objectivism
Idealism
V
Enthusiasm
Assimilation
Concept of God
LEFT
IDEAS
INNER OBJECTS
RIGHT
OUTER OBJECTS
IDEAS
DEPERSONALISATION'
REALITY
−
BODY
+
BODY
DEPERSONALISATION
REALITY

to do things they cannot account for. A case known to you all would be what happened to the Apostle Paul, namely that he was repeatedly possessed by the angel of Satan who made him utter blasphemies.[346] Complexes grow into objects that possess the individual. This also happens in the clear light of day—no trance is needed to bring it about.

These objects or figures are so-called archetypes, that is, primordial types or images.[347] For it has become apparent that these absolute objects always coincide with certain types of primitive psychology. Man has been experiencing these primordial images since time immemorial. Seen thus, Léopold is also a primordial image; only he is clothed in the form of an historical personality. Only insofar as he appears in historical garb, and hence subjectively connected with Hélène, is he obviously no longer primordial. But the fact that such a figure should appear is archetypal, and you will encounter this phenomenon time and again throughout the history of mankind across the world.

Such archetypes can appear in a broad variety of ways. For instance, they can also be equivalents of ideas. Ideas can take possession of us as if they were ghosts. "What has got into him?" we say, or "I am acting in such and such a spirit," that is, with a particular attitude that is formulated through an idea. So an idea can also be an archetype. At this stage they are no longer philosophical or religious ideas, but primordial types.

For example, primitive Negroes in the Congo area do not possess their ideas in the same way that we do. They do not have the notion of obedience or of freedom, for instance, but for them these are properties of objects. As anywhere else, there are hills, rivers, and forests in this region. Now there are certain rivers that simultaneously stand for an idea and represent it. A certain river, for example, is synonymous with freedom. You can also find this much closer to home. Just imagine you were attending an *eidgenössisches Schützenfest* [Swiss marksmen's festival],[348] listening to the official speaker holding forth: "The Alps, the *Jungfrau*,

---

[346] "And lest I should be exalted above measure through the abundance of the revelations, there was given to me a thorn in the flesh, the messenger of Satan to buffet me, lest I should be exalted above measure" (2 Corinthians 12, 7).

[347] Perhaps for reasons of easier comprehensibility, Jung is here disregarding his own distinction between archetypes *stricto sensu*, and primordial or archetypal *images* or *ideas*: "One must, for the sake of accuracy, distinguish between 'archetype' and 'archetypal ideas.' The archetype as such is a hypothetical and irrepresentable model, something like the 'pattern of behaviour' in biology" (1934b, § 6, note 8; cf. *Types*, § 513).

[348] The *Schützen* (literally, marksmen) are the old town guard, going back to medieval times; their annual festivals have continued to this day in the form of a regional patriotic tradition in Switzerland and parts of Germany. Swiss recruits keep their personal service weapon after the basic training.

this magnificent mountain, the white cross on a red background,"[349] etc., and these references are precisely such identities, meaning this and that. Someone displays a panoramic view of the Alps, for instance, and this then denotes "political freedom," or the "fatherland," or "democracy."

Matters are likewise with the natives, only much clearer. There is a hill, for instance, which stands for "loyalty towards the king," and certain plants and animals are also ascribed to it. Once a king had to deliver a speech upon his accession to office, and he had to express his joy that his chiefs should obey him, that he would not disregard their rights, and so on. Now he expressed this by simply enumerating the names of the rivers, hills, forests, and so forth; and everybody knew what was meant. He had expressed the idea. The ideas are thus always a *sous-entendu*[350] of the objects in question.

This is obviously quite galling for missionaries. One missionary sought to translate the song "Jesus, my Redeemer" into the native language. Alas, he failed to observe that the stress falls on the penultimate syllable. The natives love to sing this song. One day, the bishop visited the area. He was fluent in the native language, and discovered that they were actually singing, "Jesus, our Locust." This made perfect sense to them, since for them the idea was much better illustrated by this than by "Jesus, our Redeemer."[351]

So, archetypes were originally objects or activities. We can still find evidence for this in our language. When we say *einen Kranken behandeln* [to treat a patient],[352] this means that I request the patient to lie down in order to "treat" him with my hands. French *traitement* or English *treatment* means to "pull" or "drag" through, in other words, to extract the sickness or the evil spirit that caused it. In this context, let me mention an interesting ancient custom in England: There is a flagstone with a manhole at the center, through which I was able to just about squeeze. Stone pillars are stood to its left and right, symbolizing the beginning and

---

[349] Referring, of course, to the Swiss national flag.

[350] French, insinuation, implication.

[351] Jung gave a slightly different version of this in a question-and-answer session at the International Medical Congress for Psychotherapy at Oxford in 1938: "For years they [a primitive tribe] sing a hymn in which there is a word meaning 'hope,' or 'confidence.' A missionary who listened to that hymn didn't know the accentuation of that word properly: If you put the accent on the last syllable it means hope, and if you put it on the first, it means locust. So they sang, 'Jesus is our locust,' and that went quite well, because the locust is a religious figure in Africa" (*Jung Speaking*, pp. 112–113).

[352] Jung refers to the etymological root of the word *be-handeln* = to take something into one's hand, to "handle" something.

end.[353] One crawls through the hole in the stone in a certain direction. To this day, sick children are pulled through this hole for healing purposes. Man is "reborn" as it were, because he had not "turned out well" originally. Then one dresses him in a new shirt and new clothes, and possibly he is even given a new name, so that the illness no longer knows him. Hence, the word "treatment," from the Latin *trahere*, "to pull through." Ultimately, these matters are based upon primordial facts, primordial events, and primordial actions.

To return to my diagram, the center acts like a magnet. Whoever approaches it, is attracted and taken hold of. People who are apparently entirely concentrated in an abstract way, find themselves suddenly called back to the body by a slight pain. To an even greater extent this is true in illnesses. When people have been ill for a long time, you can clearly notice how completely their interest has become focused on themselves.

We find a similar point of attraction in section V, but there it works in the opposite direction. Here there is a magnet that draws people out of themselves, thereby inducing the phenomena of ecstasy. Here we find "the entirely other," the Latin *totaliter aliter*. Already in sections II and III we have some premonition of something "other." If somebody discovers something about himself, for example that he is a rascal, which of course can happen to the best of us, we say: but that cannot be, that is not possible. But we get a first dim inkling of something very uncanny in ourselves, something very different from the picture we had made of ourselves. Naturally we try to turn away from such an unpleasant idea, but the hunch was at work, and we were driven to see that, unbelievable as it might seem, this less reputable person is also myself. In section V this work is completed, and there appears something completely different and superhuman. When this "other" begins to draw us toward it, we forget ourselves, leading to a depersonalization, which is felt to be the "entirely other," the *totaliter aliter*. This is the sphere of ecstasy and of mystical experience, in which man is dissolved in an absolute self. The experiences told by mystics describe precisely this stepping out of reality.

Now to the right side. This is the side of consciousness, of broad daylight, of the familiar and known. It is the consciousness of the empirical and immanent I, which is usually identified with the body. In the first sphere on the right side, we first find a certainty of the corporeal I. Usually, it is as if the body could not be conscious of itself, as if it possessed

[353] This is the Mên-an-Tol monument near Madron in west Cornwall (the name means "stone with a hole" in Cornish). There exist various theories about the age and the original arrangement and purpose of the stones.

no consciousness of the I. Consciousness of the negative side exists, too. Primitives, for example, having a very limited range of consciousness, are naturally just as aware of their negative side as we are aware of our consciousness. They take it for granted that a person can at times behave one way, and at other times completely differently, that is, switch between the right and left sides of sphere I. They have no knowledge of what they were like a moment ago; they immediately forget and have no idea what they will be like in the next moment. There are people like that among us, too, people, that is, who first kick you in the belly, and then shake their head in surprise at your reaction: "Hey, what's the problem with you, I just meant you well! What on earth is the matter?"

In the second sphere, we move into the area of objects. The outer object already becomes active here. However, it is still seen completely according to one's own point of view; one is quite unable to see the object as it really is. This is the stage of subjectivism. Psychological subjectivism is a type of "autoerotism,"[354] a kind of egocentricity. Conflicts arise, since there is a clash with reality. One does not see objects and people as they are, and one does not let the "other" have its say. If I see my fellow humans only from a subjective viewpoint, I shall always see them through the veil of my consciousness. Many people founder on this, because they simply cannot escape this sphere.

The third sphere is that of personalism, where the objective factor becomes apparent, and we realize that other people have a value of their own. The veil of subjectivism is lifted, and we discover that others are not what we thought they were. Our "bêtes noires" and our treasured heroes then turn into quite ordinary people. Personalism is a state in which I realize that other people are different in a valid way; for we now no longer assume that it is wrong for them to be different from us. When someone realizes that he should never contradict his superior, precisely because he is his superior, who will otherwise fire him, he has already discovered personalism. But already in this section, where other people become real to us, a certain doubt is beginning to form with regard to the absolute reality of pure concreteness, and we are beginning to leave it behind as we move slowly towards another kind of reality.

In the fourth sphere, objectivism, we have overcome the level of commonplace personalism. In the third sphere, we had dealt with a number

[354] A term popularized by the British sexologist Havelock Ellis (1859–1939), and also adopted by Freud. In psychoanalysis, it refers "either to pleasurable activity in which the self is used as an object . . . or to a libidinal attitude, orientation, or stage of development" (Rycroft, p. 1968, 10).

of people who were admittedly different from us, for instance, more powerful, more intelligent, stupider or weaker, etc., but they appeared to us as single individuals and were unrelated. In that world, there existed a Mr. X, and a Mr. Y, etc., and society was just the companionship of companions. Society was simply a meeting of relatives, of acquaintances, or of strangers. One went bowling, attended funerals and weddings, and so on, since there were people one simply "had to" meet there.

Here, however, this personal perspective disappears and is replaced by impersonal idealism. What become important are no longer clubs and societies, composed of such and such individual members, but rather the ideas that have created these associations. The "fatherland" has ceased to consist of individuals, and has become an idea instead. A political party is no longer a personal matter. Instead, it takes on the qualities of a higher obligation, of an imperative; it is one's duty to partake, a duty, that is, in both the most ridiculous and the most sublime sense. Of course, abuse is always possible. Here the most sublime ideals can be donned just as one puts on an overcoat, by which the actual meaning is distorted. In the noblest sense, the highest virtues prevail in this sphere; but at the same time we also find the greatest vices here. After all, the *diabolica fraus*[355] can distort everything. We must never lose sight of the fact that all these general human ideas and ideals can be abused—for there is nothing in heaven or earth that the human animal will not abuse, but nevertheless real ideals do exist in this sphere, and it is these ideals that have given rise to state, society, church, religion, etc. Here we already touch upon something like a religious idea.

In section V on the right side, as in section V on the other side, the "entirely other" begins to take effect, pulling people out of themselves and their personal sphere, and bringing about a state of depersonalization. Actually, it is beyond my competence to speak about this. These poles are really beyond human understanding. A mystic or a poet occasionally reaches them, and speaks to us out of such a state of ecstasy, but any partial experience of them in our own lives is so strange to us that we may begin to doubt our mental sanity. These states cannot be ignored, however, and it is absurd to dismiss them as being "nothing but" this or that, or to say: "Well, he just wanted to play a role." When someone pays for his political convictions with his life, this cannot be shrugged off as "nothing but." These things simply do happen, but why this is so I am unable to account for, just as little as I can account for the existence of

[355] Latin, a devil's cheat.

elephants. They exist, period. It simply is a fact that people are sometimes drawn out of themselves by a power that we do not understand. We must credit the living human being for being an incredible phenomenon.

Ladies and gentlemen, this diagram is the result of much deliberation and comparison. It is the fruit of encounters with people from all walks of life, from many countries and continents, and it refers not only to the present, but also to the past. It is a diagram that one can think about very much, and for a very long time. It shows us one thing very clearly, namely, that the human I is put, as it were, between two poles. It is as if we were a core of iron placed between two powerful magnetic poles. One pulls this way, the other that way, and there is always the danger that this unity, which we call the "I," will be severed and fall into pieces. If it is split by a clear cut, so to speak, psychic dissociation or hysteria follows. If it is torn to pieces, however, we call it schizophrenia.

Lecture 15

16 February 1934

Submitted Questions

I have received some questions concerning man's position between the two poles.[356] My audience is clearly interested in my own views on the diagram discussed last time. The diagram illustrates a scale of attitudes and basic psychological forms, which represent the respective foundations for the shifts of consciousness. Now the question of the polar tension, or the opposition of the two poles, is a very difficult one. I would rather not tell you what I think about these poles, but rather explain how one arrives at such reasoning in the first place.

Obviously, consciousness cannot expand endlessly, but eventually reaches a limit where personal, human psychology ends, and something completely different begins, which we can no longer understand. If someone withdraws completely from the human sphere, our means of comprehension are exhausted. He can offer no explanation, and even if he could, we would not able to understand it. In any case it would be impossible to make any general statements about it. He might even be mad, and we would be equally unable to understand what happened to him. For this reason, I would rather not

[356] For example:

"14 February 34
Dear Doctor,
How can we explain that the Seeress of Prevorst, who lived so completely on the left side, seemed to have had such a close connection with the pole on the right side? Is this an illusion, or could the left pole be seen as the underside, so to speak, of the right pole, as on a sheet of paper? (I do not know how literally your physical analogy is to be taken.) Would it then be possible to reach one pole through the other one, or would this only be possible—psychologically speaking—by passing through the centre?

With many thanks I am
Yours respectfully,
Rita Harrvey,
eager listener in the Friday hours" (ETH Archives).

comment on these two poles, other than to state that such extreme states of consciousness do indeed exist. They are the endpoints or target points of the polar tension. They symbolize, as it were, the last realities we are able to grasp, and are negative boundary concepts, like the Kantian thing-in-itself. They are landmarks erected where someone has departed from the human sphere. We also have the testimonies of the mystics and of poets, proclaimed at the highest moments of enthusiasm, of misery and despair, and of ecstasy.

***

How does this diagram work in practice? It is not merely a speculative invention, but has arisen from practical experience, and has become a necessary basis for the sifting and ordering of the tremendously complex experiential material. What has most compelled me to devise this diagram was my direct practical experience of people and of how they think and feel. I have had to realize how incredibly different people are, although outwardly they all look similar and appear to be one large herd. Every patient is a new experience to me, because ultimately human beings are all entirely different if one probes deep enough. Naturally, such diversity leads to an enormous variety of conceptions, intentions, convictions, and so on, and that is why it is so difficult to reach a consensus.

This fact is particularly apparent in the history of psychological theories. Today, we are confronted with an enormous diversity of opinions. We have at our disposal a wide range of possibilities to explain a certain fact, each time from a completely different viewpoint. While this diversity shows how very much alive the sciences are, it also makes the clarification of empirical material all the more difficult. This is one of the reasons why the problem of psychological types, for instance, is so difficult. We must constantly argue over the words that we use. Take for example the perfectly ordinary German word *Gefühl* [feeling, sentiment]—just imagine the range of potential meanings! It holds a whole labyrinth of meaning, everybody understands it in a different way, and even the classical authors have confused it with "sensation." Most psychological terms have quite arbitrary meanings, and hardly any clear consensus exists about any of them. It's enough to drive one to despair that when I use the word *Gefühl*, for instance, everybody assumes he understands it exactly in the same way as I do. This leads to a considerable number of pitfalls that kill almost all discussion, as often witnessed at congresses of psychologists.

A recognition of this fact is one of the reasons for this attempt at clarification. Let us now see how our experiences can be incorporated into the diagram.

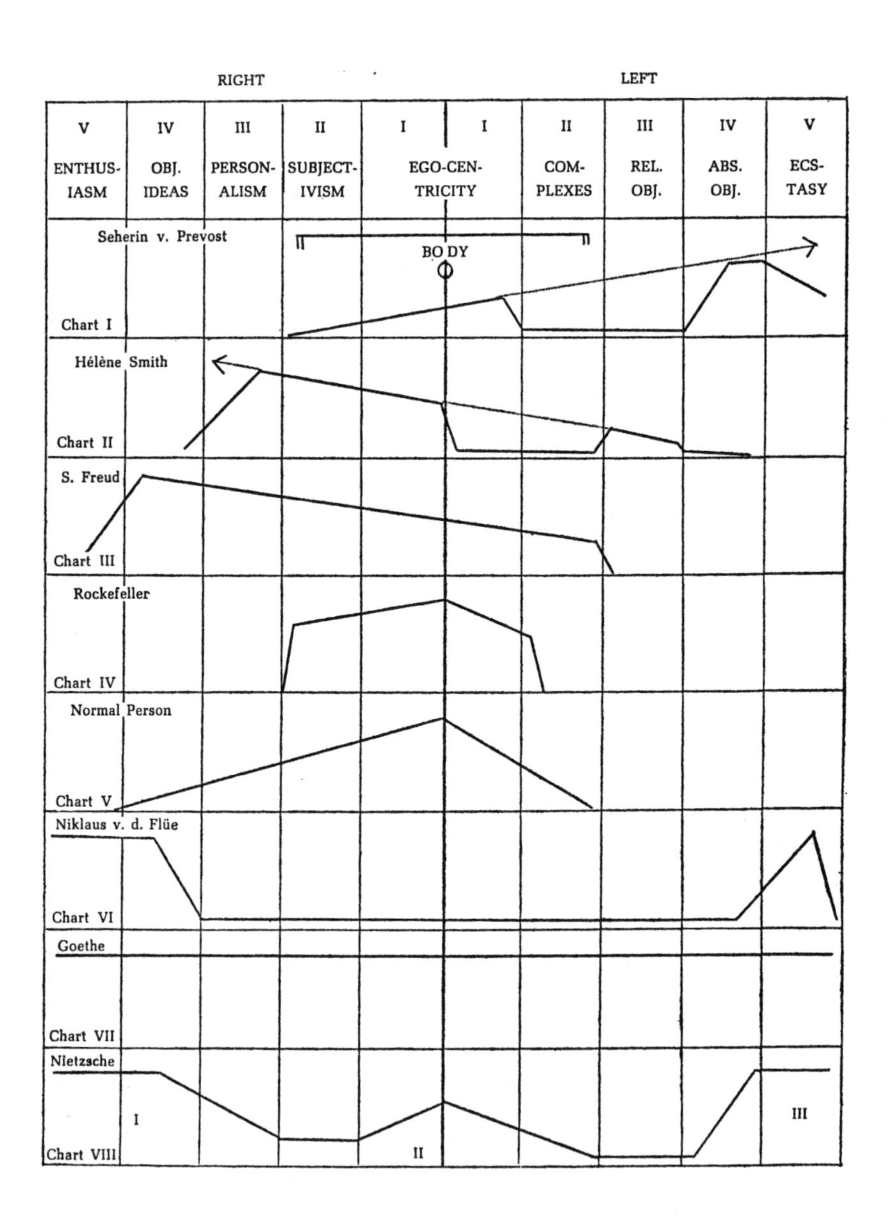

RIGHT
LEFT
V
IV
III
II
I
I
II
III
IV
V
ENTHUSIASM
OBJ. IDEAS
PERSONALISM
SUBJECTIVISM
EGO-CENTRICITY
COMPLEXES
REL. OBJ.
ABS. OBJ.
ECSTASY
Seherin v. Prevost
BODY
Chart I
Hélène Smith
Chart II
S. Freud
Chart III
Rockefeller
Chart IV
Normal Person
Chart V
Niklaus v. d. Flüe
Chart VI
Goethe
Chart VII
Nietzsche
I
II
III
Chart VIII

Let us first apply this scheme to the "Seeress of Prevorst" (Left V), and try to determine the limits of her conscious attitude. She stands very much on the left side. The main characteristic of her experience is doubtless that she lives in another reality than ours, namely in a reality made up entirely of absolute, inner objects, that is, of so-called ghosts. Thus, she reaches her highpoint at Left IV. This raises the question of how matters stand with the fifth category, in other words, to what extent is her fantasy influenced by something that is really beyond human reach? This pertains to only one point: her account of the sun-sphere. I should have to take you as far as India, China, or Tibet to prove that her experience still lies within human reach and that such things actually do exist. If I can demonstrate to a madman that his ideas do not lie beyond the sphere of the human mind, he will still feel part of human society, and there is still hope. As long as you can make yourself understood to one single person, you are not yet mad. And even if you find no such person, you should consult some old books, and perhaps there you will find something that seems familiar to you. Only when you can no longer make yourself understood will you be mad and excluded. I make every effort, therefore, when dealing with such cases, to keep the bridge of understanding open. If I can, so to speak, nod to a strange experience as to an acquaintance, the patient will be related to reality and feel reassured. If I were to say, however, "No, that is unheard of, such a thing does not exist anywhere except in your imagination," then it would be all over for the patient, and the last bridge to human relationship be broken down. The patient would be isolated in his experience, and the only door still open would lead to madness.

We do not find the Seeress in V, the sphere of ecstasy and the final fusion with what is no longer human. For her, there only exists a symbol of it, the sun-sphere. She was able to give a good description of it to her doctor, Kerner, who was really interested, and in this way she related her experience to the world. She herself did not stand in Left V, the symbol alone is there, and the book makes evident that this symbol was not very expressive and had a less profound effect on her than her experiences in the fourth category. So the line descends in V.

Now, how far does she reach into the right side? She is very much focused on her body, and prone to constant suffering; everything must revolve around her, and all she does is ponder her condition. Her consciousness, however, stretches very little further into the world of reality, and her relationship to people—for instance, to her child—is singularly subjective, quite unsteady, and largely dependent upon her mood. So she speaks to her child as if to a subjective category, and she also sees her beloved Kerner only from the perspective of the left side.

These considerations would be nothing but an idle game, were it not for the fact that we can draw certain practical conclusions from the diagram. We cannot but judge on the basis of our state of consciousness. Anyone else at exactly the same position in the chart as the Seeress will think and react in much the same way. She will never find absolute reality on the right side, but rather in her spirit-world. She will have little interest in the world of consciousness. She will pass entirely subjective judgement on outer things, will relate everything to her subjective personality, and will no longer be concerned about the outer world, even if it were on fire.

The curve also shows that the emphasis has shifted toward the left. If this situation continues to be stable, then it will remain that way; should it change, however, as in a developmental process, then the conscious attitude will shift even further towards the left side. This case exhibits one particular feature, however: I have connected the categories by simply drawing a straight line. This would mean that a conscious awareness of her personality exists in all those categories. However, this does not correspond to the facts, since there is no continuous consciousness nor, for example, any strong personification. Instead, what occurs is an eclipse of consciousness, a psychological void, followed by a somnambulistic state. The line thus falls off at Left II, and only rises again sharply at Left IV. This always indicates a split of consciousness and shows that she is an abnormal case.

Let us now have a look at the case of Hélène Smith. Here, matters are considerably different. Her figure of Léopold/Balsamo/Cagliostro is situated on the right side and very subjectively colored, thus representing a relative objectivation. There is a psychological moment behind this figure, namely, an animus, which is closely connected with her doctor and observer, Professor Flournoy. They have even engaged in a secret flirtation, although naturally under the protective cloak of secret symbols. As soon as we know this, we realize that there must be a marked emphasis on the right side, and that we cannot explain these cases in terms of the left side. We must grant her the reality of her experiences. If we attempted to explain to her that this is "nothing but" her subjective complexes, we would be wide off the mark and commit a professional blunder.

As we know, Hélène Smith was quite well adapted to life, efficient in her work, a valued sales assistant at a department store, quite sociable, etc. She was very successful in managing to interest an American, who presented her with enough dollars to provide herself with a comfortable old age; so we may conclude that she was quite clever in worldly matters.

We have also seen that her ghosts had nothing in common with those seen by the "Seeress of Prevorst." Hers are all relative and very subjective in character; they are relative objectivations of complexes. Evidently, this has to be explained from the right side. She produced neither mandalas nor other such manifestations. With her, things are only relatively genuine, although she wants them to appear genuine. They are manipulated with tremendous subtlety, but have not actually occurred. Whereas the Seeress gives us an objective picture of the left side, Hélène's information is only relatively valid; she really speaks of the right side under the guise of the left.

She liked her spirits, they brought her fame, but they remained very subjective. She would have been greatly disturbed had they become too objective and self-willed. In her case, the eclipse of consciousness begins immediately behind the I and covers most of I and II on the left side. She had no knowledge of which complexes lay behind her beloved Léopold, nor is she in any fashion open to criticism, and thus remains unaware of the shadow. If we were to show her all the things hidden in the objectivation of her Léopold, she'd be quite angry at us! According to the rule that we have established, consciousness must, if no static condition abides, tend to shift towards the right side, namely towards the objective ideals. This shift, however, is inhibited in her case by the stimulation of unconscious contents, which naturally strive for the other side.

The two women will also offer quite different stories. The Seeress will tell the whole world about her ghosts. Hélène Smith, however, will talk, behind her mask, of very tangible realities. If she is sincere, she will have a lot to say about her acquaintances, and about her Professor Flournoy; she likes to be famous, and is proud about the receptions given to her eminence at such and such a castle, etc.

Not only persons, but also theories can be viewed with the help of this diagram, including Freud's theory, which brings us to our third case. Freud's highest point lies in the sphere of objective ideals, that is, in Right IV. In fact, his idea is the only true one. When we have reached the sphere of such absolute truth, however, we have already left the human sphere, and entered the realm of the Gods. If an idea comes to possess someone so powerfully that he no longer possesses the idea but instead is possessed by it, then he has reached Right V, the level of enthusiasm.

Freud is the psychologist of the complexes, so his curve is already relatively high in Left II. Were it even higher, then this would already be neurotic, and we would have to conclude that his thinking came only out of his complexes. Now, we will refrain from assuming as much, we are much too polite for that. Is Freud somnambulistic, or does he have a blackout of consciousness anywhere? Since we have no knowledge of this, we must assume that his line is uninterrupted. Freud is keenly aware of the negative side of the unconscious. He is the psychologist of the dark side, and the human shadow is his disclosure. His fullest consciousness on the left side is in I, and he revealed his discoveries in this sphere to an astonished and shocked Europe.

Personalism occupies a strong position in his thinking. He was actually the first to discover that hysterics are in fact human beings, and not just illnesses. He taught physicians that neurotics are individuals, not merely "cases." We have to approach patients as a subjective being. Researchers like Freud must be able to be subjective; otherwise they will never be able to induce another person to open up. You can only induce the patient to declare his standpoint when you can tell him what you yourself think of him.

This is a chart where the curve of consciousness is unbroken: it runs continuously to the right side, passing through the center, but the light ends in Left II. This explains his whole attitude. He will therefore explain everything on the left side in terms of the right one, and handle everything on the left side beyond the sphere of the complexes, that is from III to V, negatively, for so-called scientific reasons. The religious phenomenon will elude him entirely, and his idea of religion will be distorted by personalism and subjectivism. His book, *The Future of an Illusion* (1927), illustrates how little Freud, despite his genius, understood the phenomenon of religion. So when someone tells him that a dream has conveyed a message to him, or brings him a vision, or when he reads of experiences of mystics and artists, he will inevitably say: "Well, that's nothing other than a complex!" He will accord it no reality. IV and V on the left side do not exist for him. God is only a complex.

Let us conclude with the case of the elder Rockefeller.[357] Here we have a very much simplified curve, with an extremely narrow consciousness, reaching its highest point in I. I once had the opportunity to speak to him,

[357] John D. Rockefeller Sr. (1839–1937), American industrialist and philanthropist; founder of the Standard Oil Company, which dominated the oil industry and was the first great U.S. business trust. His daughter Edith, and her equally rich husband Harold McCormick, were patients of Jung as well as crucial sponsors of him and the Jungian movement. Jung met Rockefeller Sr. in 1912, when he had talks with his daughter regarding a possible analysis in Zurich.

and could take a deep look into his personality, which is very complicated indeed. Rockefeller is really just a mountain of gold, and you might be wondering whether I asked him how he managed to amass such riches. But I am no longer curious about this, for I have seen that such gold is bought too dearly. The poor old gentleman is a terrible hypochondriac, and is exclusively interested in his health. He frets all day over his physical well-being, thinking of different medicines, if he should he go to the baths and, if so, to which one, thinking of trying out different diets—or also doctors! Plainly, he has a bad conscience. His secretary told me that he always carries dimes in his pockets, so that he can tip the boys who collect the balls on golf courses, or for that matter any child he chances to meet, so that it will look sweetly at him. Because he is terribly lonely.

I regret to tell you that the dear old gentleman does not move further right than to sphere II, and has never got beyond a subjective attitude. The following conversation may serve to illustrate this. I was an attentive listener, in spite of his slow speech and his long artistic pauses.

Rockefeller: Right, so you are a European. I love the Europeans, but there are some very bad people among them.
Jung: Yes, people are much the same as elsewhere, good and bad.
R.: The Austrians are very bad people.
J.: No, really, I never knew that.
(He looks at me pityingly.)
R.: Well, you don't know everything, doctor, but I expect you to realize that I am an idealist. For many years I have been striving to do something great for humanity, and to establish a standard oil price for the entire world. Every country agreed, except Austria whose government had just signed a separate treaty with Romania. The Austrians must be very bad people![358]

[358] "From 1910 to 1912, . . . the Standard Oil Trust, and . . . Austria-Hungary engaged in a bitter, protracted dispute that Austria's leading newspaper dubbed a 'Petroleum War'" (Frank, 2009, pp. 16–17; see *Neue Freie Presse*, 24 September 1910). Austria-Hungary was then the world's third-largest oil-producing country, after the USA and Russia. The anecdote about Rockefeller and the Austrians can be also found in Jung's seminar on Nietzsche's *Zarathustra* (seminar of 26 June 1935; 1988, p. 583).

Lecture 16

23 February 1934

The diagram discussed last time has prompted so many questions that I have decided to illustrate how it works with further examples.

Let as have a look at the average curve of the "normal" person. You would be ashamed to be a normal person! Schopenhauer maintains that his egotism is so great that he would even strike dead his own brother in order to grease his boots with the latter's fat.[359] Thus, the normal person is firstly very selfish and obstinate, and secondly primitive. It is a fact that the ancient cave men are still among us; you will meet them on the tram! Likewise, Neolithic men and pile dwellers. Today, we might call them imbeciles, and so on. It takes very little, and out comes the barbarian in us again. At least 70, if not 80 percent of the population still belong to the Middle Ages, so that in fact very few people are truly adjusted to the year 1934. And of these few, most have forgotten what lies behind them: that is, they have forgotten their shadow, which they carry through life behind their well-adjusted *personae* or roles. So the highest point of the curve lies in I. Normal man lives there with his body, which is an animal.

We can also assume that Right II, that is, subjectivism, assumes a high position, for the average person is extremely subjective.

In Right III, personalism, we find submission or obedience to an authority, perhaps to the *Führer*.[360] Here the curve falls off somewhat, but in recent times an intensifying seems to have occurred.

---

[359] Trying to find a "very emphatic hyperbole" for the magnitude of man's egotism, Schopenhauer came up with the following: "Many a man would be quite capable of killing another, simply to rub his boots over with the victim's fat" [*mancher Mensch wäre im Stande, einen andern todtzuschlagen, bloß um mit dessen Fette sich die Stiefel zu schmieren*]; not without adding, however: "I am only doubtful whether this, after all, is really a hyperbole" [*Aber dabei blieb mir doch der Skrupel, ob es auch wirklich eine Hyperbel sei*] (1840 [1903], p. 154; 1840 [1977], p. 238).

[360] Jung is recorded as having spoken of *the* "Führer," not *a* Führer (= leader). Hitler had been appointed chancellor the year before (30 January 1933).

In Right IV, the realm of objective ideas, the curve falls off, and consciousness has almost completely vanished; it is difficult indeed to be objective. Ideas presuppose an independence of mind and self-discipline, something that only very few people possess.

Right V is indeed very weak.

On the left side, in Section I, we find a dim idea of the dark things, but not much, then the curve sinks, consciousness extends no further, and nothing at all happens any longer.

This curve simply shows the profile of the average person, but it tells us nothing about the type. The field of the extravert lies more to the right, and that of the introvert more to the left. The latter is more conscious of his shadow, and accordingly feels somewhat inferior. He cannot meet reality directly, but has to meditate over it first. By this mechanism he avoids many pitfalls, but he also misses a great many opportunities. The extravert, on the contrary, blunders from sheer ignorance of his shadow, and is sorely handicapped when he is driven to discover his inner world.

The curve of the normal person, however, changes with the times, as collective consciousness may move to the right. With the rise of certain religious movements, when general consciousness soars, the curve can reach Right V. To cite an historical example, I refer to the wave of ecstasy that swept over the ancient world with the rise of Islam. The fanatical crying of "il Allah" is an ecstatic clamoring, which pulls man out of his instinctual, animal-like condition. In times that are more introverted, consciousness shifts more to the left, as illustrated, for example, by an interest in psychology. Such an interest shows that people have begun to become troubled in this direction (Left III), and therefore they wish to know more about this.

Now to the curve of a medieval man, Nicholas of Flüe or Brother Klaus.[361] His curve starts very high in Right V. In contrast to modern man, the life of this mystic revolved around religion, which to him was a powerful reality. He was governed by a central experience, by a spiritual power. For him, this is a conscious motif, not an illusion, but quite simply a fact. When I treat such a person, I must accept him as he is, and not as I believe he ought to be. If he were egotistical, then he would certainly

[361] Nicholas of Flüe, or Brother Klaus (1417–1487), Swiss hermit, ascetic, and visionary; canonized and declared patron saint of Switzerland by Pope Pius XII in 1947. Cf. Niklaus von Flüe, 1587. The year before, Jung had published a short treatise on him and his "so-called Trinity Vision, which was of the greatest significance for the hermit's inner life" (1933, § 477).

suppress his egotism. If I tried to explain to him that he belonged in sphere I, he would believe that I am a representative of the devil. Apart from that, Brother Klaus would of course never have consulted me for treatment in any case!

The curve falls off slightly at Right IV. Ideas played no great role in his case; he was not an educated man.

In Right III the curve sinks altogether, indicating a remarkable drive toward independence. He even left his family, and also did not shy away from pulling the highly commendable representatives of the confederate cantons by their ears at the Diet of Stans.[362]

Right I and II are practically obliterated, as was his intention and purpose.

The dark left side does not exist for him. He went his own way, unburdened by psychological problems and without pondering their background. In Left V, however, he had a powerful experience of an inner and unorthodox nature: a vision of the Holy Trinity. This left a lifelong mark on him. In the vision, a powerful face full of wrath appeared to him in a powerful apparition of light. The vision profoundly frightened Brother Klaus, and this fright impressed itself so distinctly on his face that people began to avoid him, growing frightened of him in turn. He then sought to come to terms with these experiences in a small book, and the result of his endeavors can be seen at Stans church: a painting of the vision of the Holy Trinity. He called it a vision of the Trinity, because he tried to regard it as a vision of God, so that he could bring it in line with his orthodox faith in Right V. But the terrifying, grimacing face that had appeared to him was in reality that of a *deus absconditus*.[363]

We will fly high this time and speak of Goethe. In his case, we are at a loss. With Faust, we want to exclaim, "How can I grasp you, boundless Nature?"[364] Here one actually does not know where to set the highest point. His light has shone on the whole *orbis terrarum*. In Right V, for

---

[362] In 1467, he left his wife and his ten children to lead the life of a hermit. He later returned, however, and took residence in a hermitage near his old house. At the Diet of Stans in 1481, a severe controversy arose between the five rural and the three urban cantons, which threatened to disrupt the confederation, whereupon Niklaus von Flüe, who was also a military man, was asked to mediate. On his advice, all eight cantons reached an understanding that they were to make no separate alliances of their own without the approval of a majority among the eight, which resulted in a crucial strengthening of the federal union.

[363] Latin, hidden God, the God unknowable by the human mind, as opposed to *deus revelatus*, the revealed God; after Isaiah 45, 15: "Verily thou art a God that hidest thyself, O God of Israel, the Saviour." This is a concept that played an important role in the thinking of Nikolaus von Kues, John Calvin, and Martin Luther; Jung, too, repeatedly referred to it.

[364] *Wo fass' ich dich, unendliche Natur?* (Goethe, *Faust I*, line 455)

him the face of God is revealed in nature.[365] In Left V, he fades away into an utterly unorthodox and highly original vision at the end of *Faust*. Those who think that Goethe was merely fabulating are completely off the track. He crossed the threshold in the "Dedication": "Again you come, you hovering figures."[366] Here we are in the shadow land. Seldom has anyone fathomed nature so well, and seldom has anyone seen so much of the dark world as he did. On the other hand, he could feel as happy as a pig in mud and yet suffer like a dog at the same time![367] Thus, we can safely assume that he was also at ease in sphere I. Nothing human was strange to him: *Nihil humanum [ei] alienum erat.*[368]

So we might have to draw a straight line in his case. Whether this holds really true for His Excellency, Privy Councilor Goethe, I do not know, but he was doubtless an unusually universal man. And should you take the trouble to venture into your copy of *Faust* on a Sunday you will discover Goethe's polar tension: "One impulse art thou conscious of, at best"—this is Right V—and: "O, never seek to know the other!"—Left V.[369]

Before I discuss the final curve, Nietzsche's, I should perhaps mention that such curves do not always remain valid for a person's whole life. Consciousness wanders either to the right or to the left side. Thus, it would be impossible to see Nietzsche's case as static; he is one the move. We can distinguish three phases: First, he was a man of intense spirituality and powerful ideas, so his curve reaches a highpoint in Right IV and V.

Second, a neurotic disposition begins to emerge, with its highpoint in I. This is shown by the fact that his subject and his subjectivity, though only slight at the beginning, become ever more pronounced. This is observable already at Right III and Right II through the strong emergence

---

[365] As in the *Prologue in Heaven* (ibid.).

[366] *Ihr naht euch wieder, schwankende Gestalten* (ibid., the opening line of the play). In Goethe's writings, *schwankend* (literally shaking, unsteady, wavering) describes a figure that has not yet assumed a definite form (editorial note in Goethe, 1996, p. 505).

[367] MS: *dass es einem so sauwohl und so sauwehe zugleich sein könne*. Probably an allusion to Goethe's entry in his Swiss diary of summer, 1775: *Dass es der Erde so sauwohl und so weh ist zugleich!* [That earth should feel so happy as a pig in mud and so wretched at one and the same moment!] Cf. also *Faust*: "We feel so bloody jolly, just like five hundred hogs!" [*Uns ist ganz kannibalisch wohl, / Als wie fünfhundert Säuen!*] (lines 2293–2294).

[368] Latin, nothing that is human was alien (to him). A variation of Terence's (195/185–159 BCE) dictum: *Humani nihil a me alienum puto* (in *Heauton Timorumenos*) (often also quoted as *nihil humanum mihi alienum [esse] puto*).

[369] Faust's disciple Wagner sings the praises of book learning, to which Faust replies: "One instinct are you conscious of, at best; Oh, never should you know the other! Two souls, alas, reside within my breast, And the one wants to separate from the other" [*Du bist dir nur des einen Triebs bewusst; / O lerne nie den andern kennen! / Zwei Seelen wohnen ach in meiner Brust, / Die eine will sich von der andern trennen*] (lines 1110–1117).

of the I. His subject emerges ever more prominently: subjectivism. Left II is also very pronounced. His neurosis tended to elevate the curve further and further, even in a peculiar tendency toward a psychology of complexes. Nietzsche was thus a precursor of analytical psychology, since he was very much preoccupied with complexes. We find nothing at all in Left III.

It is not until Left IV that again a sudden increase occurs. A highpoint is reached at Left IV und V. Coming from the side of medieval man, he reached, passing through an incredible phase of conflict, the side of modern man, in which process a tremendous Dionysian experience occurred in V: "I, the last disciple and initiate of the God Dionysos."[370] Had he been static, this would have led to somnambulism, to an eclipse of consciousness. Instead he had his Dionysian experience, an unorthodox, authochthonous experience, which then also became efficacious in the figure of Zarathustra. "Then one turned to two, and Zarathustra passed me by."[371] Nietzsche is not playing some literary trick here. Someone *did* confront him, and he did experience him.

What we have here is a tremendous tension between the two poles. To say it with a simile: looking from the right side we see a house from the outside, and looking from the left side we see the same house from within. The same holds true for persons; there is an incredible discrepancy between how they look from the outside and from the inside, and it is a veritable art to guess from the outside how it looks from within. Many highly interesting stories could be told about this, but unfortunately a lack of time prevents me from doing so now.

Let us return once again to the original diagram on which the last eight charts were based. That scale is best imagined spatially, as a plain on which circles are drawn for the sections. Now put yourself into the center of these circles, thus defining back, front, right, and left. Imagine

---

[370] *[I]ch, der letzte Jünger und Eingeweihte des Gottes Dionysos* (Nietzsche, 1886, p. 238).

[371] Nietzsche described the appearance of Zarathustra in a little poem, written in Sils-Maria in the Engadine: "I sat here, waiting, waiting, but for nothing; beyond good and evil, savoring now the light, and now the shade; it was all game, all lake, all midday, it was just time without a goal. Then suddenly, dear friend!, one turned to two, and Zarathustra passed me by" [*Hier sass ich, wartend, wartend,—doch auf Nichts, / Jenseits von Gut und Böse, bald des Lichts / Geniessend, bald des Schattens, ganz nur Spiel, / Ganz See, ganz Mittag, ganz Zeit ohne Ziel. // Da, plötzlich, Freundin! wurde Eins zu Zwei—/ Und Zarathustra gieng an mir vorbei . . .*] (in Nietzsche, 1882, appendix; KSA 3, p. 649). Cf. Jung, 1988 [1934–1839], p. 744; 2014 [1936–1941], pp. 174–175; and 1934b, §§ 77–78, where he also quotes the last lines and again stresses: "Zarathustra is more for Nietzsche than a poetic figure; he is an involuntary confession, a testament."

that the air is very thick, foggy, so that you cannot see very far. Visibility ends at a certain point, and you must now advance to see what lies there.

Throughout, the center remains the fixed or starting point, that is, the primitive consciousness of one's own body, and of one's instinctiveness. Thereafter comes the second sphere on the right side, in which we are still under the spell of our own subjectivity. In III, we encounter people who are different from us, and who might make a deep impression on us, and to whom we might feel inferior or superior. Then comes a region, IV, where people become what they represent and become somehow elevated to a more or less superhuman level by their functions. Somebody is then no longer Mr. Jones, for example, but has become General Jones.

We find the same situation on the left side. We encounter the great difficulty, however, that our contemporary consciousness is oriented one-sidedly, that is, to the right side. Too few people are conscious of the left side, because for most this is unchartered territory. It is as if we believed that only Europe existed. Now imagine a visitor from the United States, who tells us about the country and its customs, about New York and its skyscrapers: we would either believe that he had been dreaming or was simply telling an amusing story—or even that he was mad.

It is extremely rare that someone is willing to abandon the present position of his consciousness. Once consciousness has claimed a certain resting point, it can barely be removed from its place. It creates convictions, and people get so stuck in them that anything different is just seen as bad. Therefore there is always mass murder when a new idea comes into the world.

When someone reaches his highpoint in Right III, that is, where humanity is still regarded as a gathering of more or less distinct individuals with whom I have dealings, then he will be inclined to assume that this is the way of the world, and that nothing else exists. Or if I stand close to the center, in I or II, I will be so convinced of my reality that no other argument can top my experience of, for instance, "This hurts!" and nothing else will reach me. The sun is shining outside, the birds are singing, and it is a wonderful world—but, alas, "I have a headache!" Or someone is so fascinated by his own ideas that everything that does not fit into this particular world is taken to be inferior, or has even been especially invented by the devil to torment the good; and everything else will be exterminated with fire and sword. Or if someone is complex-ridden, then he will perceive also others colored in this way.

Such viewpoints set in stone enormously exacerbate matters. Whoever stands at the far right and beholds the entire world on his left will think it a sorry mistake. A lieutenant colonel in the medical corps once said to me: "All psychoanalysts should have their skulls cracked!" If someone comes with a new idea, he must fear catastrophic consequences and the cracking of his skull. Hence all the difficulties at psychological congresses—precisely on account of such immovable viewpoints. It really requires a catastrophe or a severe neurosis to dislocate people.

When one studies such a person, however, it becomes apparent that viewed from his own standpoint he is in his right. He sits on his throne, and chooses not to step down from it until it collapses. Generally, this is experienced as a catastrophe. Every point of view has an inner logic, and is a reality. What we can do is to persuade the other, with more or less cunning, to descend from his throne and to view the world from that small hill over there for a change. We must be able to abandon our point of view, make a *sacrificium intellectus*,[372] and also a sacrifice of our morals, of our notion of right and wrong, since there is a consensus that the "other" exists, too. All those countries that we have not yet discovered exist, even though we have not discovered them!

An intuitive type, it is true, sees dozens of possibilities in other spheres, but he does not actually go there to experience them. For example, he sees a person living in Right IV as he appears to him from his vantage point in Left III. Consequently, the intuitive may see a great deal of which the man in Right IV is not aware, but what the intuitive says is unintelligible to the other man [in Right IV] because he does not know that Left III exists at all. As America existed before it was discovered, so it is with the dark areas of the human soul. They are forever present and at work; it is merely a question of whether we notice this. There are a great number of "protective" mechanisms that prevent us from noticing our dark side. But others may have noticed it. People sometimes move over to another point of view for a short time and then slip back to their former little hill. If you suggest, "But you said you saw such and such," they reply, "Did I say such a thing? . . . I have forgotten . . . how strange!" It is as if you had made a faux pas.

Continuous progression in this circle marks a development of the range of consciousness. It constitutes a shifting of viewpoint, which as such is closely associated with the maturing of the personality. We do not know why consciousness moves sometimes to the right or to the left. It does not do so in all cases. Mrs. Hauffe, for instance, had remained on the

---

[372] Latin, sacrifice of the intellect; the third sacrifice demanded by the founder of the Jesuits, St. Ignatius Loyola.

same side since she had been a child; it was her congenital temperament. There are dispositions that a priori localize consciousness, like a magnet. Sometimes it is a matter of blows of fate, such as great catastrophes or major disappointments, of which the legends of the saints provide some excellent examples.

One question, then, would be whether Mrs. Hauffe, whose relationship with the left pole was particularly close, had also contact with the right pole? Whether a mysterious connection exists between the two poles must be answered in the affirmative. It is as if we were looking at the house at the same time from within and from without. That a person might be able realize both poles, however, might be possible in theory, but I would doubt very much whether it is possible in practice. As we have seen in Nietzsche's case, for instance, this is a painful transition from one side to the other. The poles exert an overwhelming pull, so that it is impossible to be situated at both poles simultaneously. You are either inside the house, or outside.

As to the practical use of the diagram to classify writers, and so forth, this is certainly possible, but only in those cases that have been subjected to a very thorough psychological study. Above all, we must know what these individuals are conscious of and what they are not. We must not mess around with this diagram!

Well, after having impressed you with this description of the powerful tension in the human soul, one could almost believe that I believe in a secret dualism, as if souls were stretched between two relentless poles that are never able to come together. This is the case, true, and yet it also is not; for where there is a separating force, a unifying one will arise. Now this diagram refers exclusively to the shifts of consciousness, to its localization. But this does not tell us anything about the quality of the personality that is the bearer of this consciousness.

The case of Hélène Smith would provide some clues, with the help of which I could have explained to you how consciousness can alter without changing its location as it were. It is as if amid the polar tension a new consciousness suddenly emerges at the center from this animalistic I, which, as it were, unfolds in cycles, and develops into a different kind of consciousness. We call this characteristic function, which occurs naturally in every polar tension and seeks to unite the opposites of our nature, the transcendent function.[373]

[373] On the concept of the transcendent function see note 329.

Bibliography

Anquetil-Duperron, Abraham Hyacinthe (1801, 1802). *Oupnek'hat (id est, Secretum tegendum): opus ipsa in India rarissimum, continens antiquam et arcanam, seu theologicam et philosophicam, doctrinam, e quatuor sacris Indorum Libris, Rak Beid, Djedjr Beid, Sam Beid, Athrban Beid, excerptam; ad verbum, è Persico idiomate, Samskreticis vocabulis intermixto, in Latinum conversum; dissertationibus et annotationibus, difficiliora explanantibus, illustratum*; studio et opera Anquetil Duperron. Vols. 1 and 2. Strassbourg: Levrault, 9, 10.

Baldwin, James Mark (1913). *History of Psychology: A Sketch and an Interpretation.* London: Watts.

Benoit, Pierre (1919). *L'Atlantide.* Paris: Albin Michel. An English translation (*Atlandida*) is available at various Internet sites, e.g., Project Gutenberg.

Bergson, Henri (1915). "La philosophie française." *Revue de Paris*, May 15, 1915: 236–256.

Berkeley, (Bishop) George (1710). *A Treatise Concerning the Principles of Human Knowledge.* Dublin: printed by Aaron Rhames for Jeremy Pepyat.

Béroalde de Verville: see Colonna, Francesco.

Binet, Alfred (1892). *Les Alterations de la Personnalité.* Paris: Librairie Félix Alcan (Bibliothèque scientifique internationale, publiée sous la direction de M. Ém. Alglave). *Alterations of Personality.* Trans. Helen Green Baldwin, with notes and preface by J. Mark Baldwin. New York: D. Appleton & Company, 1896.

Bishop, Paul (1997). "The Descent of Zarathustra and the Rabbits: Jung's Correspondence with Elisabeth Förster-Nietzsche." *Harvest, Journal for Jungian Studies*, 43(1): 108–123.

Bonnet, Charles (1760). *Essai analytique sur les facultés de l'âme.* Copenhagen: Cl. and Ant. Philibert.

Boring, Edwin G. (1929). *A History of Experimental Psychology.* New York: Century.

Brachfeld, Oliver (1954). Gelenkte Tagträume als Hilfsmittel der Psychotherapie. *Zeitschrift für Psychotherapie*, 4: 79–93.

Büchner, Ludwig (1855). *Kraft und Stoff. Empirisch-naturphilosophische Studien. In allgemein-verständlicher Darstellung.* Frankfurt am Main: Meidinger. 6th edition: Frankfurt am Main: Meidinger, 1859. 19th edition: Leipzig: Theodor Thomas, 1898. *Force and Matter: Empirico-Philosophical Studies, Intelligibly Rendered.* Ed. J. Frederick Collingwood. London: Trübner & Co., 1864.

Burckhardt, Jacob (1860). *Die Cultur der Renaissance in Italien. Ein Versuch.* Leipzig: E. A. Seeman, 2nd edition, 1869. *The Civilisation of the Renaissance in Italy*. Trans. S. G. C. Middlemore. London: S. Sonnenschein; New York: Macmillan & Co., 1904.

Burnham, John (ed.) (2012). *After Freud Left: Centennial Reflections on His 1909 Visit to the United States*. Chicago: University of Chicago Press.

Carus, Carl Gustav (1846). *Psyche. Zur Entwicklungsgeschichte der Seele*. Pforzheim: Flammer und Hoffmann. *Psyche: On the Development of the Soul*. Intr. James Hillman. Trans. R. Welch. Zurich: Spring Publications, 1970; 2nd ed. Dallas: Spring Publications, 1989.

Carus, Carl Gustav (1866). *Vergleichende Psychologie oder Geschichte der Seele in der Reihenfolge der Thierwelt*. Vienna: Wilhelm Braumüller.

Colonna, Francesco [?] (1499 [1467]). *Poliphili hypnerotomachia, ubi humana omnia non nisi somnium esse ostendit, atque obiter plurima scitu sanequam digna commemorat*. Venice: Aldus Manutius. *The Strife of Love in a Dream*. 1592. Béroalde de Verville, François (1600). *Le tableau des riches inventions, couvertes du voile des feintes amoureuses, qui sont representees dans le Songe de Poliphile*. Desvoilees des ombres du Songe, & subtilement exposees par Beroalde De Verville. Paris: Matthieu Guillemot. *Hypnerotomachia Poliphili: The Strife of Love in a Dream*. Trans. Jocelyn Godwin. London: Thames & Hudson, 1999.

Condillac, Étienne Bonnot de (1754). *Traité des sensations*. 2 vols. London, Paris: de Bure, 1754. *Traité des sensations*. New and expanded ed. 2 vols. London, Paris: Barrois aîné, 1788. *Traité des sensations*. In *Oeuvres philosophiques de l'abbé de Condillac*. New ed. Parma, Paris: Guillemard, 1792. *Traité des sensations*. In *Oeuvres de Condillac*. Revues, corrigées par l'Auteur, imprimées sur ses manuscrits autographes, et augmentées de "La Langue des Calculs", ouvrage posthume. Vol. 3. Paris: Ch. Houel, 1798. *Des Herrn Abbt's* [*sic*] *Condillac Abhandlung über die Empfindungen*. Trans. Joseph Maria Weissegger von Weisseneck. Vienna: Johann David Hörling, 1791. Can be downloaded as pdf file from the site of the Bayerische Staatsbibliothek. *Condillac's Abhandlung über die Empfindungen*. Aus dem Französischen übersetzt mit Erläuterungen und einem Excurs über das binoculare Sehen von Dr. Eduard Johnson. Leipzig: Verlag der Dürr'schen Buchhandlung, 1870. *Abhandlung über die Empfindungen*. Ed. Lothar Kreimendahl. Hamburg: Meiner, 1983.

D'Alembert, Jean-Baptiste le Rond (1751). *Discours préliminaire de l'Encyclopédie*. Paris: chez Briasson, David, Le Breton, Durant. *Preliminary Discourse to the Encyclopedia of Diderot*. Trans. Richard N. Schwab. Chicago: The University of Chicago Press, 1995.

Descartes, René (1641). *Meditationes de prima philosophia, in qua Dei existentia et animæ immortalitas demonstratur*. Paris: Apud Michaelem Soly, via Iacobea, sub signo Phoenicis. *Meditations on First Philosophy*. With selections from *The Objections* and *Replies*. Ed. John Cottingham. Cambridge: Cambridge University Press, 1986; rev. ed. 1996.

Dessoir, Max (1902). *Geschichte der neueren deutschen Psychologie. 1*. Vol. 2. Thoroughly rev. ed. Berlin: Carl Duncker.

Dessoir, Max (1911). *Abriß einer Geschichte der Psychologie*. Heidelberg: Carl Winter's Universitätsbuchhandlung.

Dumas, Alexandre (père) (1846–1848). *Mémoires d'un Médecin, Joseph Balsamo*. Serial in *La Presse*, 31 May 1846, to 4 January 1848. *Mémoires d'un Médecin, Joseph Balsamo. I.* New ed. Paris: Michel Lévy Frères, 1860. *Joseph Balsamo* and *The Memoirs of a Physician* (2 vols.). In *The Works of Alexandre Dumas in Thirty Volumes*. Vols. 6 and 7. New York: P. F. Collier, 1902.

Dunne, John William (1927). *An Experiment with Time*. London: A. & C. Black.

Ellenberger, Henri (1970). *The Discovery of the Unconscious*. New York: Basic Books.

Falzeder, Ernst (2016). "Types of Truth: Jung's Philosophical Roots." *Jung Journal, Culture & Psyche*, 10(3): 14–30.

Faust, Johann (1501). *D. Faustus Magus Maximus Kundlingensis Original Dreyfacher Höllenzwang*. Reprint: *Faust's dreifacher Höllenzwang: Dr. Faust's Magia naturalis et innaturalis*. Berlin: Schikowski, 2002.

Fechner, Gustav Theodor [writing under the pseudonym of Dr. Mises] (1836). *Das Büchlein vom Leben nach dem Tode*. Leipzig: Insel-Verlag. *On Life After Death*. London: Sampson Low, Marston, Searle & Rivington, 1882. *The Little Book of Life After Death*. Trans. Mary C. Wadsworth. With a preface by William James. Boston: Little, Brown and Company, 1904, 1912. *On Life After Death*. Trans. Hugo Wernekke. Chicago: The Open Court Publishing Company, 1906. *Life After Death*. New York: Pantheon, 1943. "The Little Book of Life after Death." *Journal of Pastoral Counseling: An Annual*, 1992, 27: 7–31.

Fechner, Gustav Theodor (1848). *Nanna oder das Seelenleben der Pflanzen* [*Nanna, or the Soul-Life of Plants*]. Leipzig: Leopold Voß.

Fechner, Gustav Theodor (1851). *Zend-Avesta oder Über die Dinge des Himmels und des Jenseits vom Standpunkt der Naturbetrachtung*. 3 vols. Leipzig: Leopold Voß. Reprint Karben: Petra Wald, 1992.

Fechner, Gustav Theodor (1860). *Elemente der Psychophysik*. 2 vols. Leipzig: Breitkopf & Härtel. *Elements of Psychophysics*. Ed. David H. Howes and Edwin G. Boring. Trans. Helmut E. Adler. New York: Holt, Rinehart and Winston, 1966.

Fechner, Gustav Theodor (1861). *Über die Seelenfrage. Ein Gang durch die sichtbare Welt, um die unsichtbare zu finden*. Leipzig: C. F. Amelang's Verlag.

Fechner, Gustav Theodor (1879). *Die Tagesansicht gegenüber der Nachtansicht*, Leipzig: Breitkopf und Härtel. Reprint of the 2nd edition of 1904. Karben: Petra Wald, 1994.

Fierz-David, Linda (1947). *Der Liebestraum des Poliphilo. Ein Beitrag zur Psychologie der Renaissance und der Moderne*. Zurich: Rhein-Verlag. *The Dream of Poliphilo*. Trans. Mary Hottinger. New York: Pantheon, 1950.

Flournoy, Théodore (1900 [1899]). *Des Indes à la planète Mars. Étude sur un cas de somnambulisme avec glossolalie*. Paris, Geneva: F. Alcan, Ch. Eggimann. *Die Seherin von Genf*. Foreword by Max Dessoir. Authorised trans. Leipzig: Felix Meiner Verlag, 1914. *From India to the Planet Mars: A Case of Multiple Personality with Imaginary Languages*. Foreword by C. G. Jung and commentary by Mireille Cifali. Trans. Daniel B. Vermilye. Ed. and introduced by Sonu Shamdasani. Princeton: Princeton University Press, 1994.

France, Anatole (1908). *L'Île des Pingouins*. Paris: Calmann-Lévy. *Penguin Island*. Trans. A. W. Evans. New York: Blue Ribbon Books, 1909. Reprint New York, Berlin: Mondial, 2005.

Francke, K. B. (1878). *Die Psychologie und Erkenntnislehre des Arnobius*. Dissertation. Leipzig.

Frank, Alison (2009). "The Petroleum War of 1910: Standard Oil, Austria, and the Limits of the Multinational Corporation." *The American Historical Review*, 114(1): 16–41.

Freud, Sigmund (1900 [1899]). *Die Traumdeutung*. Vienna, Leipzig: Franz Deuticke. In *Gesammelte Werke*. Frankfurt am Main: S. Fischer, 1942ff., Vol. 2/3. *The Interpretation of Dreams*. In *The Standard Edition of the Complete Psychological Works of Sigmund Freud*. London: The Hogarth Press and The Institute of Psycho-Analysis, 1953ff. [= SE], vols. 4 and 5.

Freud, Sigmund (1914). *On the History of the Psycho-Analytic Movement*. SE 14.

Freud, Sigmund (1927). *The Future of an Illusion*. SE 21.

Frobenius, Leo (1904). *Das Zeitalter des Sonnengottes*. Berlin: Georg Reimer.

Gedächtnisausstellung Otto Meyer(-Amden). Kunsthaus Zürich, 22. Dez. 1933 bis 28. Jan. 1934. Vollst. Verzeichnis m. 16 Abb. (1933). Zurich: Kunsthaus Zürich.

Goethe, Johann Wolfgang von (1996). *Werke*. Vol. 3, *Dramatische Dichtungen I. Hamburger Ausgabe*. Ed. Erich Trunz. Munich: Deutscher Taschenbuch Verlag, 16th revised edition.

Grau, Kurt Joachim (1922). *Bewusstsein, Unbewusstes, Unterbewusstes*. Munich: Rösl.

Gulyga, Arsenij V. (1987). *Immanuel Kant: His Life and Thought*. Basel, Boston: Birkhäuser.

Hall, G. Stanley (1912). *Founders of Modern Psychology*. New York: Appleton and Co.

Hartley, David (1749). *Observations on Man, his Frame, his Duty, and his Expectations*. Printed by S. Richardson; For James Leake and Wm. Frederick, Booksellers in Bath; And sold by Charles Hitch and Stephen Austen, Booksellers in London.

Hartmann, Eduard von (1869). *Philosophie des Unbewussten. Versuch einer Weltanschauung. Speculative Resultate nach inductiv-naturwissenschaftlicher Methode*. Berlin: Carl Duncker. *The Philosophy of the Unconscious: Speculative Results According to the Inductive Method of Physical Science*. Trans. William C. Coupland. London: Kegan Paul, Trench & Trübner, 1931.

Hartmann, Eduard von (1901). *Die moderne Psychologie. Eine kritische Geschichte der deutschen Psychologie in der zweiten Hälfte des neunzehnten Jahrhunderts*. Leipzig: Hermann Haacke.

Heidelberger, Michael (1993). *Die innere Seite der Natur. Gustav Theodor Fechners wissenschaftlich-philosophische Weltauffassung*. Frankfurt am Main: Vittorio Klostermann GmbH. *Nature from Within. Gustav Theodor Fechner and His Psychophysical Worldview*. Trans. Cynthia Klohr. Pittsburgh: University of Pittsburgh Press, 2004.

Heidelberger, Michael (2010). "Gustav Theodor Fechner and the Unconscious." In Nicholls & Liebscher, 2010, pp. 200–240.

Herbart, Johann Friedrich (1816). *Lehrbuch zur Psychologie*. Konigsberg: August Wilhelm Unzer. In *Joh. Fr. Herbart's sämtliche Werke in chronologischer Reihenfolge* (1887–1912). Vol. 4, *A Textbook in Psychology*. Trans. M. K. Smith. New York: Appleton, 1891.

Herbart, Johann Friedrich (1824–1825). *Psychologie als Wissenschaft, neu gegründet auf Erfahrung, Metaphysik und Mathematik*. Konigsberg: August Wilhelm Unzer. In *Joh. Fr. Herbart's sämtliche Werke in chronologischer Reihenfolge* (1887–1912). Part 1 in Vol. 5, pp. 177–434; Part 2 in Vol. 6, pp. 1–340.

Herbart, Johann Friedrich (1839–1840). *Psychologische Untersuchungen*. 2 vols. Gottingen: Druck und Verlag der Dieterichschen Buchhandlung.

Herbart, Johann Friedrich (1887–1912). *Joh. Fr. Herbart's sämtliche Werke in chronologischer Reihenfolge*. Vols. 1–11 ed. Karl Kehrbach, Vols. 12–19 ed. † Karl Kehrbach und Otto Flügel. Vols. 1 and 2. Leipzig: Veit & Comp.; Vols. 3–19: Langensalza: Hermann Beyer und Söhne. 2nd reprint of the edition of 1887–1912: Aalen: Scientia-Verlag, 1989.

Hesse, Hermann (2006 [1916–1944). *"Die dunkle und wilde Seite der Seele." Briefwechsel mit seinem Psychoanalytiker Josef Bernhard Lang 1916–1944*. Ed. Thomas Feitknecht. Frankfurt am Main: Suhrkamp.

Hume, David (1739–1740). *A Treatise of Human Nature, Being an Attempt to Introduce the Experimental Method of Reasoning into Moral Subjects*. Ed. L. A. Selby-Bigge. 2nd ed. revised by P. H. Nidditch. Oxford: Clarendon Press, 1975.

Hume, David (1748). *An Enquiry Concerning Human Understanding*. Reprints, among others: *An Enquiry Concerning Human Understanding, with A Letter from a Gentleman to His Friend in Edinburgh and Hume's Abstract of A Treatise of Human Nature*. Ed. Eric Steinberg. Indianapolis: Hackett, 2nd ed. 1993. *An Enquiry Concerning Human Understanding*. Mineola, NY: Dover Publications, 2004. Ed. Peter Millican. Oxford: Oxford University Press: 2008.

Hyslop, James Hervey (1905). *Science and a Future Life*. Boston: H. B. Turner & Co., 1905; London: G. P. Putnam's Sons, 1906. Available in various reprints and e-book formats.

James, W. (1886). "Report of the Committee on Mediumistic Phenomena." *Proceedings of the American Society for Psychical Research*, 1: 102–106.

James, William (1890a). *The Principles of Psychology*. 2 vols. New York: Henry Holt & Co.; London: Macmillan & Co.

James, William (1890b). "A Record of Observations of Certain Phenomena of Trance: Part III." *Proceedings of the Society for Psychical Research*, 6: 651–659.

James, William (1909). "Report on Mrs. Piper's Hodgson Control." *Proceedings of the American Society for Psychical Research*, 3: 470–589.

Janet, Pierre (1889). *L'Automatisme psychologique: Essai de psychologie expérimentale sur les formes inférieures de la vie mentale*. Paris: Félix Alcan. Reprint Société Pierre-Janet, 1973.

Janet, Pierre (1903). *Les obsessions et la psychasthénie*. 2 vols. Paris: Félix Alcan.

Janet, Pierre (1919). *Les médications psychologiques*. Paris: Alcan. *Psychological Healing: A Historical and Clinical Study*. London: Allen and Unwin, 1925.

Janet, Pierre (1930). "Autobiography of Pierre Janet." In Carl Murchison (ed.), *History of Psychology in Autobiography*. Vol. 1. Worcester, MA: Clark University Press, pp. 123–133.

Janz, Curt Paul (1978). *Friedrich Nietzsche. Biographie*. Vol. 1. Munich: Deutscher Taschenbuch Verlag, 1981.

Jarrett, James L. (1981). "Schopenhauer and Jung." *Spring: An Annual of Archetypal Psychology*, 193–204.

Jones, Ernest (1953). *The Life and Work of Sigmund Freud*. Vol. 1, *The Formative Years and the Great Discoveries, 1856–1900*. New York: Basic Books.

Jung, C. G. (1902). *On the Psychology and Pathology of So-Called Occult Phenomena*. *CW* 1, pp. 3–88.

Jung, C. G. (1905a). "Cryptomnesia." *CW* 1, pp. 95–108.

Jung, C. G. (1905b). "On Spiritualistic Phenomena. *CW* 18, pp. 293–308.

Jung, C. G. (1906). "The Psychological Diagnosis of Evidence." *CW* 2, pp. 318–352.

Jung, C. G. (1911/1912). *Wandlungen und Symbole der Libido. Jahrbuch für psychoanalytische und psychopathologische Forschungen*, 1911, 3(1): 120–227; 1912, 4(1): 162–464. In book form: Leipzig: Deuticke, 1912. Reprint: Munich: Deutscher Taschenbuch Verlag, 1991. In revised form (1950) and under new title, *Symbole der Wandlung*, in *GW* 5; *Symbols of Transformation*, in *CW* 5.

Jung, C. G. (1913 [1911]). "On the Doctrine of Complexes." *CW* 2, pp. 598–604.

Jung, C. G. (1917–1942). *Die Psychologie der unbewussten Prozesse. Ein Überblick über die moderne Theorie und Methode der analytischen Psychologie*. Zurich: Rascher. In revised form and under new title, *On the Psychology of the Unconscious*, in *CW* 7, pp. 3–122.

Jung, C. G. (1919). "Instinct and the Unconscious." *CW* 8, pp. 129–138.

Jung, C. G. (1921). *Psychological Types*. *CW* 6.

Jung, C. G. (1926). "Spirit and Life." *CW* 8, pp. 319–337.

Jung, C. G. (1926 [1924]). *Analytical Psychology and Education: Three Lectures*. *CW* 17, pp. 63–132.

Jung, C. G. (1927 [1931]). "The Structure of the Psyche." *CW* 8, pp. 139–158.

Jung, C. G. (1928). *On Psychic Energy*. *CW* 8, pp. 3–66.

Jung, C. G. (1929). "Commentary on 'The Secret of the Golden Flower.'" *CW* 13, pp. 1–56.

Jung, C. G. (1930a). "Introduction to Kranefeldt's 'Secret Ways of the Mind.'" *CW* 4, pp. 324–332.

Jung, C. G. (1930b). "Psychology and Literature." *CW* 15, pp. 84–108.

Jung, C. G. (1931a). "Basic Postulates of Analytical Psychology." *CW* 8, pp. 338–357.

Jung, C. G. (1931b). "Archaic Man." *CW* 10, pp. 50–73.

Jung, C. G. (1932). "The Hypothesis of the Collective Unconscious." *CW* 18, pp. 515–516.

Jung, C. G. (1933). "Brother Klaus." *CW* 11, pp. 316–323.

Jung, C. G. (1934a). "A Review of the Complex Theory." *CW* 8, pp. 92–104.

Jung, C. G. (1934b). "Archetypes of the Collective Unconscious. *CW* 9/1, pp. 3–41.

Jung, C. G. (1934 [1968]). Letter to *Neue Zürcher Zeitung*, 15 March 1934. *CW* 10, § 1934, fn. 544.
Jung, C. G. (1935a). "Foreword to Mehlich, 'I. H. Fichtes Seelenlehre und ihre Beziehung zur Gegenwart. '". *CW* 18, pp. 770–772.
Jung, C. G. (1935b). "Psychological Commentary on *The Tibetan Book of the Dead*. *CW* 11, pp. 509–528.
Jung, C. G. (1935/36 [1943]). *Psychology and Alchemy*. *CW* 12.
Jung, C. G. (1936 [1937]). "Psychological Factors Determining Human Behavior." *CW* 8, pp. 114–125.
Jung, C. G. (1936/1937). "The Concept of the Collective Unconscious." *CW* 9/1, pp. 42–53.
Jung, C. G. (1938 [1954]). "Psychological Aspects of the Mother Archetype." *CW* 9/1, pp. 75–110.
Jung, C. G. (1939a). "Conscious, Unconscious, and Individuation." *CW* 9/1, pp. 275–289.
Jung, C. G. (1939b). "Concerning Rebirth." *CW* 9/1, pp. 113–147.
Jung, C. G. (1939 [1937]). *Psychology and Religion*. The Terry Lectures. *CW* 11, pp. 3–105.
Jung, C. G. (1940). "The Psychology of the Child Archetype." *CW* 9/1, pp. 151–181.
Jung, C. G. (1941). "Transformation Symbolism in the Mass." *CW* 11, pp. 201–296.
Jung, C. G. (1943 [1942]). "Psychotherapy and a Philosophy of Life." *CW* 16, pp. 76–83.
Jung, C. G. (1945a). "The Phenomenology of the Spirit in Fairytales." *CW* 9/1, pp. 207–253.
Jung, C. G. (1945b). "The Philosophical Tree." *CW* 13, pp. 251–350.
Jung, C. G. (1946a). *The Psychology of the Transference*. *CW* 16, pp. 163–326.
Jung, C. G. (1946b). *On the Nature of the Psyche*. *CW* 8, pp. 159–234.
Jung, C. G. (1947[1946]). "Foreword to Fierz-David, 'Der Liebestraum des Poliphilo.'" *CW* 18, pp. 780–781.
Jung, C. G. (1950a). "Foreword to Moser, 'Spuk: Irrglaube oder Wahrglaube? '" *CW* 18, pp. 317–320.
Jung, C. G. (1950b). "Concerning Mandala Symbolism." *CW* 9/1, pp. 355–384.
Jung, C. G. (1950 [1929]). "Freud and Jung: Contrasts." *CW* 4, pp. 333–340.
Jung, C. G. (1951). *Aion: Researches into the Phenomenology of the Self*. *CW* 9/2.
Jung, C. G. (1952). "Synchronicity: An Acausal Connecting Principle." *CW* 8, pp. 417–531.
Jung, C. G. (1954 [1939]). "Psychological commentary on *The Tibetan Book of the Great Liberation*." *CW* 11, pp. 475–508.
Jung, C. G. (1955). Appendix: Mandalas. *CW* 9/1, pp. 387–390.
Jung, C. G. (1955/1956). *Mysterium Coniunctionis. An Inquiry Into the Separation and Synthesis of Psychic Opposites in Alchemy.* *CW* 14.
Jung, C. G. (1957 [1916]). "The Transcendent Function." *CW* 8, pp. 67–91.
Jung, C. G. (1959). "Introduction to Toni Wolff's 'Studies in Jungian Psychology.'" *CW* 10, pp. 469–476.

Jung, C. G. (1962). *Memories, Dreams, Reflections*. Recorded and edited by Aniela Jaffé. Trans. Richard and Clara Winston. London: Fontana Press, 1995.

Jung, C. G. (1963 [1959]). "Foreword to Brunner, 'Die Anima als Schicksalsproblem des Mannes.'" *CW* 18, pp. 543–547.

Jung, C. G. (1964 [1961]). "Symbols and the Interpretation of Dreams." *CW* 18, pp. 183–264.

Jung, C. G. (1973). *Letters I, 1906–1950*. Ed. Gerhard Adler in collaboration with Aniela Jaffé. Trans. R. F. C. Hull. Bollingen Series XCV. Princeton: Princeton University Press.

Jung, C. G. (1975). *Letters II, 1951–1961*. Ed. Gerhard Adler in collaboration with Aniela Jaffé. Trans. R. F. C. Hull. Bollingen Series XCV. Princeton: Princeton University Press.

Jung, C. G. (1983 [2000]). *The Zofingia Lectures*. With an Introduction by Marie-Louise von Franz. Ed. William McGuire. Trans. Jan van Heurck. Princeton: Princeton University Press. Reprint London: Routledge, 2000.

Jung, C. G. (1984). *Dream Analysis: Notes of the Seminar Given in 1928–1930*. Ed. William McGuire. Princeton: Princeton University Press.

Jung, C. G. (1987 [2008]). *Children's Dreams: Notes from the Seminar Given in 1936–1940*. Ed. Lorenz Jung and Maria Meyer-Grass. Trans. Ernst Falzeder. Philemon Series. Princeton: Princeton University Press, 2008.

Jung, C. G. (1988 [1934–1939]). *Nietzsche's Zarathustra: Notes of the Seminar Given in 1934–1939*. Ed. James L. Jarrett. 2 vols. Princeton: Princeton University Press.

Jung, C. G. (2012 [1925]). *Introduction to Jungian Psychology: Notes of the Seminar Given by Jung on Analytical Psychology in 1925*. Ed. Sonu Shamdasani. Philemon Series. Princeton: Princeton University Press.

Jung, C. G. (2012 [1912]). *Jung contra Freud: The 1912 New York Lectures on the Theory of Psychoanalysis*. Intro. Sonu Shamdasani. Trans. R. F. C. Hull. Bollingen Series. Princeton: Princeton University Press.

Jung, C. G. (2014 [1936–1941]). *Dream Interpretation Ancient & Modern: Notes from the Seminar Given in 1936–1941*. Ed. John Peck, Lorenz Jung, and Maria Meyer-Grass. Trans. Ernst Falzeder. Philemon Series. Princeton: Princeton University Press.

Jung, C. G. (forthcoming). *The German Seminar of 1931*. Philemon Series.

Jung, C. G. and Hans Schmid-Guisan (2013). *The Question of Psychological Types: The Correspondence of C. G. Jung and Hans Schmid-Guisan, 1915–1916*. Ed. John Beebe and Ernst Falzeder. Trans. Ernst Falzeder. Philemon Series. Princeton: Princeton University Press.

Kant, Immanuel (1766). *Träume eines Geistersehers, erläutert durch Träume der Metaphysik*. Konigsberg: Johann Jacob Kanter. In *Immanuel Kants Sämmtliche Werke*. Ed. Karl Rosenkranz and Fried. Wilh. Schubert. Part 7, section 1, *Kleine anthropologisch-praktische Schriften*. Ed. Friedrich Wilhelm Schubert. Leipzig: Leopold Voss, 1838. Reprint, critical-textual ed. and with appendices by Rudolf Malter. Stuttgart: Philipp Reclam jun., 1976, 2008. *Dreams of a Spirit-Seer, Illustrated by Dreams of Metaphysics*. Ed. Frank Sewall. Trans. F. Goerwitz. London: Swan Sonnenschein & Co.; New York: Macmillan Co., 1900.

Kant, Immanuel (1781/1787). *Critik der reinen Vernunft*. Riga: Johann Friedrich Hartknoch. *The Critique of Pure Reason*. Trans. Paul Guyer and Allen Wood. Cambridge: Cambridge University Press, 1998.

Kant, Immanuel (1783). *Prolegomena zu einer jeden künftigen Metaphysik die als Wissenschaft wird auftreten können*. Riga: Johann Friedrich Hartknoch. Reprint Stuttgart: Philipp Reclam jun., 1989. *Kant's Prolegomena to Any Future Metaphysics*. Trans. Paul Carus. Chicago: Open Court Publishing Company, 1902, 3rd ed. 1912.

Kant, Immanuel (1786). *Metaphysische Anfangsgründe der Naturwissenschaft*. Riga: Johann Friedrich Hartknoch. *Metaphysical Foundations of Natural Science*. Trans. Ernest Belfort Bax. 2nd revised ed. London: George Bell and Sons, 1891.

Kant, Immanuel (1798). *Anthropologie in pragmatischer Hinsicht*. Konigsberg: Friedrich Nicolovius. *Anthropology from a Pragmatic Point of View*. Trans. and ed. Robert B. Louden. Cambridge: Cambridge University Press, 2006.

Kant, Immanuel (1902/1910 ff.). *Kants gesammelte Schriften*. Ed. Königlich-Preußische Akademie der Wissenschaften, Vols. 1–22; ed. Deutsche Akademie der Wissenschaften, Vols. 22–31. Berlin.

Kerner, Justinus Andreas Christian (1829). *Die Seherin von Prevorst. Eröffnungen über das innere Leben und über das Hineinragen einer Geisterwelt in die unsere*. 2 vols. Stuttgart, Tubingen: J. G. Cotta'sche Buchhandlung, expanded and revised ed. Stuttgart, Tubingen: J. G. Cotta'scher Verlag. Reprint: Kiel: J. F. Steinkopf Verlag, 2012. *The Seeress of Prevorst, Being Revelations Concerning the Inner-Life of Man, and the Inter-Diffusion of a World of Spirits in the One We Inhabit*. Trans. Catherine Crowe. London: J. C. Moore, 1845. Digital reprint: Cambridge: Cambridge University Press, 2011.

Kerner, Justinus Andreas Christian (1831–1839). *Blätter aus Prevorst. Originalien und Lesefrüchte für Freunde des innern Lebens*. Communicated by the editor of *Die Seherin von Prevorst. 1.-12. Sammlung*. Vols. 1–7: Karlsruhe: Verlag von Gottlieb Braun; vols. 8–12: Stuttgart: Fr. Brodhag'sche Buchhandlung.

Kerner, Justinus Andreas Christian (1856). *Franz Anton Mesmer aus Schwaben, Entdecker des thierischen Magnetismus. Erinnerungen an denselben, nebst Nachrichten von den letzten Jahren seines Lebens zu Meersburg am Bodensee*. Frankfurt: Literarische Anstalt.

Kipling, Rudyard (1894). *The Jungle Book*. London: Macmillan & Co.

Kipling, Rudyard (1895). *The Second Jungle Book*. London: Macmillan & Co.

Kipling, Rudyard (1902). *Just So Stories for Little Children*. London: Macmillan and Co.

Kitcher, Patricia (1990). *Kant's Transcendental Psychology*. New York: Oxford University Press.

Klemm, Otto (ed.) (1934). *Bericht über den XIII. Kongress der Deutschen Gesellschaft für Psychologie in Leipzig vom 16.-19. Oktober 1933*. Jena: Gustav Fischer.

Knigge, Adolph Freiherr (1788). *Über den Umgang mit Menschen*. 2 vols. Hannover: Schmidt'sche Buchhandlung. *Practical Philosophy of Social Life, or The Art of Conversing with Men*. Trans. P. Will. Lansingburgh, NY: Penniman & Bliss, 1805.

La Mettrie, Julien Offray de (1748). *L'homme machine*. Leiden: Elie Luzac Fils.

Lamprecht, Karl Gotthard (1905). *What Is History? Five Lectures on the Modern Science of History*. Trans. Ethan Allen Andrews. New York, London: Macmillan & Co.

Lavater, Johann Caspar (1775–1778). *Physiognomische Fragmente zur Beförderung der Menschenkenntnis und Menschenliebe*. 4 vols. Leipzig; Winterthur: Weidmanns Erben und Reich; Heinrich Steiner und Compagnie.

Leibniz, Gottfried Wilhelm (1684). *Meditations on Knowledge, Truth and Ideas*. Trans. Jonathan Bennett, at http://ebookbrowse.com/leibniz-meditations-on-knowledge-truth-and-ideas-pdf-d84731352. In *Die philosophischen Schriften von Gottfried Wilhelm Leibniz* (1875–1890). Ed. C. I. Gerhardt. 7 vols. Berlin: Weidmannsche Buchhandlung. Volume 4, pp. 422–426.

Leibniz, Gottfried Wilhelm (1981 [1704–06]). *New Essays on Human Understanding*. Trans. Peter Remnant and Jonathan Bennett. Cambridge: Cambridge University Press.

Leibniz, Gottfried Wilhelm & Christian Wolff (1860). *Briefwechsel zwischen Leibniz und Christian Wolff: Aus den Handschriften der Koeniglichen Bibliothek zu Hannover*. Ed. Carl Immanuel Gerhardt. Halle: H. W. Schmidt.

Lévy-Bruhl (1910). *Les fonctions mentales dans les sociétés inférieures*. Paris: F. Alcan.

Liebscher, Martin (2018). "Schopenhauer und Carl Gustav Jung." In Daniel Schubbe and Martin Koßler (eds.). *Schopenhauer-Handbuch. Leben—Werk—Wirkung*. Stuttgart: J. B. Metzler, pp. 325–329.

Lütkehaus, Ludger (ed.) (2005). *Dieses wahre innere Afrika. Texte zur Entdeckung des Unbewussten vor Freud*. Gießen: Psychosozial Verlag.

Maine de Biran (1834–1841). *Oeuvres philosophiques de Maine de Biran*. Published by Victor Cousin. Paris: Librairie de Ladrange.

Maine de Biran (1859). *Oeuvres inédites de Maine de Biran*. Published by Ernest Naville, with the collaboration of Marc Debrit. Paris: Dezobry, E. Magdeleine & Co.

Maury, Louis Ferdinand Alfred (1861). *Le sommeil et les rêves. Études psychologiques sur ces phénomènes et les divers états qui s'y rattachent, suivies de recherches sur le développement de l'instinct et de l'intelligence avec le phénomène du sommeil*. Paris: Didier.

McGuire, William & R. F. C. Hull (eds.) (1977). *C. G. Jung Speaking. Interviews and Encounters*. Bollingen Series XCVII. Princeton: Princeton University Press.

Mead, George Robert Stow (1919). *The Doctrine of the Subtle Body in Western Tradition: An Outline of what the Philosophers Thought and Christians Taught on the Subject*. London: J. M. Watkins.

Morgenstern, Christian (1905). *Galgenlieder*. Berlin: Cassirer. *Galgenlieder*. Ed. Joseph Kiermeier-Debre. Munich: Deutscher Taschenbuch-Verlag (dtv Bibliothek der Erstausgaben) 1998. *Gallows Songs*. Trans. W. D. Snodgrass and Lore Segal. Ann Arbor: University of Michigan Press, 1967. *Songs from the Gallows: Galgenlieder*. Trans. Walter Arndt. New Haven: Yale University Press, 1993.

Morris, Earl Halstead (1931). *The Temple of the Warriors: The Adventure of Exploring and Restoring a Masterpiece of Native American Architecture in the*

Ruined Maya City of Chichen Itzá, Yucatan. 2 vols. New York: C. Scribner's Sons; Washington: Carnegie Institution of Washington; reprint: New York: AMS Press, 1980.

Murchison, Carl (ed.) (1925). *Psychologies of 1925*. Worcester, MA: Clark University Press.

Nagy, Marilyn (1991). *Philosophical Issues in the Psychology of C. G. Jung*. Albany: State University of New York Press.

Nicholls, Angus & Martin Liebscher (eds.) (2010). *Thinking the Unconscious: Nineteenth-Century German Thought*. Cambridge: Cambridge University Press.

Nietzsche, Friedrich (1882). *Die fröhliche Wissenschaft*. Chemnitz: Ernst Schmeitzner. *Kritische Studienausgabe in 15 Bänden*. Ed. Gorgio Colli and Mazzino Montinari. Berlin, New York: de Gruyter, 1980, 1988 [= KSA], Vol. 3. *The Gay Science: With a Prelude in Rhymes and an Appendix of Songs*. Trans. with commentary by Walter Kaufmann. New York: Vintage Books, 1974.

Nietzsche, Friedrich (1883–1885). *Also sprach Zarathustra. Ein Buch für alle und keinen*. Chemnitz: Ernst Schmeitzner, 1883 (parts 1 and 2); Chemnitz: Ernst Schmeitzner, 1884 (part 3); 1885 (part 4, privately printed); Leipzig: Constantin Georg Naumann, 1892 (first complete edition, ed. Peter Gast, a pseudonym for Heinrich Köselitz). KSA, Vol. 4. *Thus Spoke Zarathustra: A Book for Everyone and No One*. Trans. R. J. Hollingdale. London: Penguin Classics, 2003.

Nietzsche, Friedrich (1886). *Jenseits von Gut und Böse. Vorspiel einer Philosophie der Zukunft*. Leipzig: Constantin Georg Naumann. KSA, Vol. 5. *Beyond Good and Evil: Prelude to a Philosophy of the Future*. Ed. Rolf-Peter Horstmann and Judith Norman. Trans. Judith Norman. Cambridge: Cambridge University Press, 2002.

Niklaus von (der) Flüe (1587). *Zwey-und-neuntzig Betrachtung und Gebett deß Gottseligen fast andächtigen Einsidels Bruder Clausen von Underwalden: sambt seinen Lehren, Sprüchen und Weissagungen, von seinem Thun und Wesen: neben angehängtem kurtzen Bericht, was von ihme einmal zuhalten sey; auch von zweyen andern namhafften und seligen Einsideln S. Beat und S. Meynrat: Durch den Ehrwürdigen und Hochgelehrten Herrn D. Petrum Canisium d. Societes Jesu Theologum, von newem corrigiert und gebessert*. An jetzo zum andern mal im Truck außgangen. Ingolstatt: 1587.

Platner, Ernst (1776). *Philosophische Aphorismen nebst einigen Anleitungen zur philosophischen Geschichte*. Leipzig: Schwickertscher Verlag.

Proclus Lycaeus (n.d.). *Proclus' Commentary on the "Timaeus" of Plato*. Trans. Thomas Taylor. Frome, Somerset: Prometheus Trust, 1988, 1998.

Reid, Thomas (1846). *The Works of Thomas Reid, D.D., Now Fully Collected, with Selections from His Unpublished Letters*. Preface, notes, and supplementary dissertations by Sir William Hamilton. Edinburgh: MacLachlan, Stewart and Co.; London: Longman, Brown, Green & Longmans.

Ribot, Theódule (1870). *La psychologie anglaise contemporaine (École expérimentale)*. Paris: Librairie philosophique de Ladrange.

Ribot, Théodule (1879). *La psychologie allemande contemporaine (École expérimentale)*. Paris: Librairie Germer Baillière et Cie.

Rider Haggard, Henry (1886–87). *She: A History of Adventure. The Graphic*: 20 October 1886; 8 January 1887 (2 parts). Book version: London: Longmans, Green & Co., 1887.

Röhricht, Alexander (1893). *Die Seelenlehre des Arnobius, nach ihren Quellen und ihrer Entstehung untersucht. Ein Beitrag zum Verständnis der späteren Apologetik der alten Kirche*. Hamburg: Agentur des Rauhen Hauses.

Rosenzweig, Saul (1992). *Freud, Jung and Hall the King-Maker. The Historic Expedition to America (1909), with G. Stanley Hall as Host and William James as Guest*. St. Louis: Rana House Press.

Rousseau, Jean-Jacques (1755 [1754]). *Discours sur l'origine et les fondements de l'inégalité parmi les hommes*. Amsterdam: Marc Michel Rey.[374]

Rousseau, Jean-Jacques (1761). *Julie, ou la nouvelle Héloïse*. Amsterdam: Marc Michael Rey.

Rousseau, Jean-Jacques (1762a). *Du contrat social ou Principes du droit politique*. Amsterdam: Marc Michel Rey.

Rousseau, Jean-Jacques (1762b). *Émile, ou De l'éducation*. The Hague: Jean Néaulme.

Rousseau, Jean-Jacques (1782–1789). *Les confessions*. Geneva: [no publisher].

Rycroft, Charles (1968). *A Critical Dictionary of Psychoanalysis*. Harmondsworth: Penguin, 1972.

Rykwert, Joseph (1988). *The Idea of a Town: The Anthropology of Urban Form in Rome, Italy and the Ancient World*. Cambridge: MIT Press.

Sage, Michael (1904). *Mrs. Piper and the Society for Psychical Research*. Trans. and slightly abridged, Noralie Robertson. With a Preface by Sir Oliver Lodge. New York: Scott-Thaw Co.

Schelling, Friedrich Wilhelm Joseph (1800). *System des transcendentalen Idealismus*. Tubingen: J. G. Cotta'sche Buchhandlung. *System of Transcendental Idealism*. Trans. Peter Heath. With an Introduction by Michael Vater. Charlottesville: University of Virginia Press, 1993.

Schopenhauer, Arthur (1819). *Die Welt als Wille und Vorstellung*. Vol. 1. Leipzig: Bibliographisches Institut F. A. Brockhaus. 2nd expanded ed., 1844; 3rd rev. and expanded ed., 1859. *Zürcher Ausgabe, Werke in zehn Bänden*. Zurich: Diogenes, 1977. Vols. 1 and 2. *The World as Will and Idea*. Trans. R. B. Haldane and J. Kemp. Boston: Ticknor & Co., 3rd ed. 1887. London: Kegan Paul, Trench, Trübner & Co., 7th ed. ca. 1909. *The World as Will and Representation*. Vols. 1 and 2. Trans. E. F. J. Payne. New York: Dover, 1969.

Schopenhauer, Arthur (1836). *Ueber den Willen in der Natur. Eine Erörterung der Bestätigungen, welche die Philosophie des Verfassers, seit ihrem Auftreten, durch die empirischen Wissenschaften erhalten hat*. Frankfurt am Main: Siegmund Schmeber. 3rd expanded and revised ed., ed. Julius Frauenstädt, Leipzig: F. A. Brockhaus, 1867. Zürcher Ausgabe, Werke in zehn Bänden. Zurich: Diogenes, 1977. Vol. 5., *On the Will in Nature*. Trans. Madame Karl Hillebrand. Scotts Valley, CA: CreateSpace (an Amazon company), 2010.

[374] For English editions of Rousseau's works see *The Collected Writings of Rousseau*. Ed. C. Kelley, R. Masters, and P. Stillman. London: University of New England Press, 1990ff.

Schopenhauer, Arthur (1840). *Preisschrift über die Grundlage der Moral*. Zürcher Ausgabe, Werke in zehn Bänden. Zurich: Diogenes, 1977. Vol. 6. *The Basis of Morality*. Trans. Arthur Brodrick Bullock. London: Swan Sonnenschein & Co., 1903.

Schopenhauer, Arthur (1844). *Die Welt als Wille und Vorstellung*. Vol 2. Leipzig: Bibliographisches Institut F. A. Brockhaus. Zürcher Ausgabe, Werke in zehn Bänden. Zurich: Diogenes, 1977. Vols. 3 and 4. *The World as Will and Idea*. Vol. 3, *Containing Supplements to Part of the Second Book and to the Third and Fourth Books*. Trans. R. B. Haldane and J. Kemp. London: Kegan Paul, Trench, Trübner & Co., 6th ed. 1909. *The World as Will and Representation*. Vols. 1 and 2. Trans. E. F. J. Payne. New York: Dover, 1969.

Schopenhauer, Arthur (1851). *Parerga und Paralipomena: kleine philosophische Schriften*. 2 vols. Berlin: A. W. Hayn. *Zürcher Ausgabe, Werke in zehn Bänden*. Zurich: Diogenes, 1977. Vols. 9 and 10. *Parerga and Paralipomena*. Trans. E. F. J. Payne. Oxford: Clarendon Press, 2000. *Parerga and Paralipomena: A Collection of Philosophical Essays*. Trans. T. Bailey Saunders. New York: Cosimo, 2007.

Schopenhauer, Arthur (1966–75[1985]). *Der handschriftliche Nachlaß*. Vol. 3, *Berliner Manuskripte (1818–1830)*. Ed. Arthur Hübscher. Munich: Deutscher Taschenbuch Verlag, 1985.

Sechzig Upanishad's des Veda (1897). Trans. Paul Deussen. Leipzig: F. A. Brockhaus. 2nd ed. 1905. *Upanishaden: Die Geheimlehre des Veda*. Ed. Peter Michel, trans. Paul Deussen. Wiesbaden: Marixverlag, 2nd ed. 2006.

Shamdasani, Sonu (2000). Unpublished notes to a presentation given in San Francisco.

Shamdasani, Sonu (2003). *Jung and the Making of Modern Psychology: The Dream of a Science*. Cambridge: Cambridge University Press.

Shamdasani, Sonu (2012). *C. G. Jung: A Biography in Books*. New York, London: W. W. Norton & Company.

Shamdasani, Sonu (2015). "'S. W.' and C. G. Jung: Mediumship, Psychiatry and Serial Exemplarity." *History of Psychiatry*, 26: 288–302.

Stewart, Dugal (1792). *Elements of the Philosophy of the Human Mind. Volume 1*. In *The Works of Dugal Stewart in Seven Volumes*, Volume 1. Cambridge: Hilliard and Brown, 1829.

Sturm, Thomas (2001). "How not to Investigate the Human Mind: Kant and the Impossibility of Empirical Psychology." In Eric Watkins (ed.), *Kant and the Sciences*. New York, Oxford: Oxford University Press.

Swedenborg, Emanuel (1758). *De coelo et ejus mirabilibus et de inferno, ex auditis et visis*. London: [s. n.]. *Heaven and Its Wonders and Hell from Things Heard and Seen*. Trans. George F. Dole. West Chester, PA: Swedenborg Foundation, 2001.

Taylor, Eugene (1980). "Jung and William James." *Spring, A Journal of Archetype and Culture*, 20: 157–169.

Tetens, Johannes Nikolaus (1777). *Philosophische Versuche über die menschliche Natur und ihre Entwickelung*. Leipzig: M. G. Weidmanns Erben und Reich.

Thomas Aquinas (1256–1259). *Quaestiones disputatae de veritate*. Ed. A. Dondaine. Editio Leonina. Rome: Editori di san Tommaso, 1972–1976.

Wehr, Gerhard (1972). *C. G. Jung und Rudolf Steiner, Konfrontation und Synopse*. Stuttgart: Klett-Cotta. Various editions and reprints (e.g., Zürich: Diogenes, 1996).

Wolff, Christian (1719–1720). *Vernünftige Gedancken von Gott, der Welt und der Seele des Menschen, auch allen Dingen überhaupt, den Liebhabern der Wahrheit mitgetheilet von Christian Wolffen*. Frankfurt am Main, Leipzig.

Wolff, Christian (1721). *Oratorio de Sinarum philosophia practica / Rede über die praktische Philosophie der Chinesen* [On the practical philosophy of the Chinese]. Reprint ed. Michael Albrecht. Hamburg: Felix Meiner, 1985.

Wolff, Christian (1728). *Philosophia rationalis sive logica methodo scientifica pertractata et ad usum scientiarum atque vitae aptata. Discursus praeliminaris de philosophia in genere*. Reprint of 3rd edition (1740). In Günter Gawlick and Lothar Kreimendahl (eds.), *Einleitende Abhandlung über Philosophie im allgemeinen (Discursus praeliminaris)*. Stuttgart: Frommann-Holzboog, 2006.

Wolff, Christian (1732). *Psychologia empirica methodo scientifica pertractata*. Frankfurt a. M., Leipzig: Renger.

Wolff, Christian (1733). *Der vernünftigen Gedancken von Gott, der Welt und der Seele des Menschen, auch allen Dingen überhaupt, anderer Teil, bestehend in ausführlichen Anmerckungen, zu besserem Verstande und bequemerem Gebrauche derselben heraus gegeben von Christian Wolffen*. Frankfurt am Main: Joh. Benj. Andreä & Heinr. Hort.

Wolff, Christian (1962 ff.). *Gesammelte Werke*. Ed. Jean École et al. Hildesheim, Zurich, New York: Olms.

Zweig, Arnulf (1967). "Gustav Theodor Fechner“. In Paul Edwards [Eisenstein] (ed.), *The Encyclopedia of Philosophy*. Vol. 3. New York: Macmillan.

Index

The Collected Works of C. G. Jung

Editors: Sir Herbert Read, Michael Fordham, and Gerhard Adler; executive editor, William McGuire. Translated by R.F.C. Hull, except where noted.

1. PSYCHIATRIC STUDIES (1957; 2d ed., 1970)
   On the Psychology and Pathology of So-Called Occult Phenomena (1902)
   On Hysterical Misreading (1904) Cryptomnesia (1905)
   On Manic Mood Disorder (1903)
   A Case of Hysterical Stupor in a Prisoner in Detention (1902)
   On Simulated Insanity (1903)
   A Medical Opinion on a Case of Simulated Insanity (1904)
   A Third and Final Opinion on Two Contradictory Psychiatric Diagnoses (1906)
   On the Psychological Diagnosis of Facts (1905)

2. EXPERIMENTAL RESEARCHES (1973)
   Translated by Leopold Stein in collaboration with Diana Riviere
   STUDIES IN WORD ASSOCIATION 1904–7, 1910)
   The Associations of Normal Subjects (by Jung and F. Riklin) An Analysis of the Associations of an Epileptic
   The Reaction-Time Ratio in the Association Experiment
   Experimental Observations on the Faculty of Memory
   Psychoanalysis and Association Experiments
   The Psychological Diagnosis of Evidence
   Association, Dream, and Hysterical Symptom
   The Psychopathological Significance of the Association Experiment
   Disturbances in Reproduction in the Association Experiment
   The Association Method

Prefaces to "Collected Papers on Analytical Psychology" (1916, 1917)
The Significance of the Father in the Destiny of the Individual (1909/1949)
Introduction to Kranefeldt's "Secret Ways of the Mind" (1930)
Freud and Jung: Contrasts (1929)

5. SYMBOLS OF TRANSFORMATION
([1911–12/1952] 1956; 2d ed., 1967)

PART I

Introduction
Two Kinds of Thinking
The Miller Fantasies: Anamnesis
The Hymn of Creation
The Song of the Moth

PART II

Introduction
The Concept of Libido
The Transformation of Libido
The Origin of the Hero
Symbols of the Mother and Rebirth
The Battle for Deliverance from the Mother
The Dual Mother
The Sacrifice
Epilogue
Appendix: The Miller Fantasies

6. PSYCHOLOGICAL TYPES ([1921] 1971)
A revision by R.F.C. Hull of the translation by H. G. Baynes
Introduction
The Problem of Types in the History of Classical and Medieval Thought
Schiller's Idea on the Type Problem
The Apollonian and the Dionysian
The Type Problem in Human Character
The Type Problem in Poetry
The Type Problem in Psychopathology
The Type Problem in Aesthetics
The Type Problem in Modern Philosophy
The Type Problem in Biography
General Description of the Types
Definitions
Epilogue

Four Papers on the Psychological Typology (1913, 1925, 1931, 1936)

7. TWO ESSAYS ON ANALYTICAL PSYCHOLOGY (1953; 2d ed., 1966)

On the Psychology of the Unconscious (1917/1926/1943)
The Relations between the Ego and the Unconscious (1928)
Appendix: New Paths in Psychology (1912); The Structure of the Unconscious (1916) (new versions, with variants, 1966)

8. THE STRUCTURE AND DYNAMICS OF THE PSYCHE (1960; 2d ed., 1969)

On Psychic Energy (1928)
The Transcendent Function ([1916] 1957)
A Review of the Complex Theory (1934)
The Significance of Constitution and Heredity and Psychology (1929)
Psychological Factors Determining Human Behavior (1937)
Instinct and the Unconscious (1919)
The Structure of the Psyche (1927/1931)
On the Nature of the Psyche (1947/1954)
General Aspects of Dream Psychology (1916/1948)
On the Nature of Dreams (1945/1948)
The Psychological Foundations of Belief in Spirits (1920/1948)
Spirit and Life (1926)
Basic Postulates of Analytical Psychology (1931)
Analytical Psychology and *Weltanschauung* (1928/1931)
The Real and the Surreal (1933)
The Stages of Life (1930–31) The Soul and Death (1934)
Synchronicity: An Acausal Connecting Principle (1952)
Appendix: On Synchronicity (1951)

9. PART I. THE ARCHETYPES AND THE COLLECTIVE UNCONSCIOUS (1959; 2d ed., 1968)

Archetypes of the Collective Unconscious (1934/1954)
The Concept of the Collective Unconscious (1936)
Concerning the Archetypes, with Special Reference to the Anima Concept (1936/1954)
Psychological Aspects of the Mother Archetype (1938/1954)
Concerning Rebirth (1940/1950)
The Psychology of the Child Archetype (1940)
The Psychological Aspects of the Kore (1941)

The Phenomenology of the Spirit in Fairytales (1945/1948)
On the Psychology of the Trickster-Figure (1954)
Conscious, Unconscious, and Individuation (1939)
A Study in the Process of Individuation (1934/1950)
Concerning Mandala Symbolism (1950)
Appendix: Mandalas (1955)

9. PART II. AION ([1951] 1959; 2d ed., 1968)
RESEARCHES INTO THE PHENOMENOLOGY OF THE SELF
the Ego
The Shadow
The Syzygy: Anima and Animus
The Self
Christ, a Symbol of the Self
The Signs of the Fishes
The Prophecies of Nostradamus
The Historical Significance of the Fish
The Ambivalence of the Fish Symbol
The Fish in Alchemy
The Alchemical Interpretation of the Fish
Background to the Psychology of Christian Alchemical Symbolism
Gnostic Symbols of the Self
The Structure and Dynamics of the Self Conclusion

10. CIVILIZATION IN TRANSITION (1964; 2d ed., 1970)
The Role of the Unconscious (1918)
Mind and Earth (1927/1931) Archaic Man (1931)
The Spiritual Problem of Modern Man (1928/1931)
The Love Problem of a Student (1928)
Woman in Europe (1927)
The Meaning of Psychology for Modern Man (1933/1934)
The State of Psychotherapy Today (1934)
Preface and Epilogue to “Essays on Contemporary Events” (1946)
Wotan (1936)
After the Catastrophe (1945)
The Fight with the Shadow (1946)
The Undiscovered Self (Present and Future) (1957)
Flying Saucers: A Modern Myth (1958)
A Psychological View of Conscience (1958)
Good and Evil in Analytical Psychology (1959)

Introduction to Wolff's "Studies in Jungian Psychology" (1959)
The Swiss Line in the European Spectrum (1928)
Reviews of Keyserling's "America Set Free" (1930) and "La Révolution Mondiale" (1934)
The Complications of American Psychology (1930)
The Dreamlike World of India (1939)
What India Can Teach Us (1939)
Appendix: Documents (1933–38)

11. PSYCHOLOGY AND RELIGION: WEST AND EAST (1958; 2d ed., 1969)

WESTERN RELIGION

Psychology and Religion (the Terry Lectures) (1938/1940)
A Psychological Approach to Dogma of the Trinity (1942/1948)
Transformation Symbolism in the Mass (1942/1954)
Forewords to White's "God and the Unconscious" and Werblowsky's "Lucifer and Prometheus" (1952)
Brother Klaus (1933)
Psychotherapists or the Clergy (1932)
Psychoanalysis and the Cure of Souls (1928)
Answer to Job (1952)

EASTERN RELIGION

Psychological Commentaries on "The Tibetan Book of Great Liberation" (1939/1954) and "The Tibetan Book of the Dead" (1935/1953)
Yoga and the West (1936)
Foreword to Suzuki's "Introduction to Zen Buddhism" (1939)
The Psychology of Eastern Meditation (1943)
The Holy Men of India: Introduction to Zimmer's "Der Weg zum Selbst" (1944)
Foreword to the "I Ching" (1950)

12. PSYCHOLOGY AND ALCHEMY ([1944] 1953; 2d ed., 1968)

Prefatory Note to the English Edition ([1951?] added 1967)
Introduction to the Religious and Psychological Problems of Alchemy
Individual Dream Symbolism in Relation to Alchemy (1936)
Religious Ideas in Alchemy (1937)
Epilogue

13. ALCHEMICAL STUDIES (1968)

Commentary on "The Secret of the Golden Flower" (1929)
The Visions of Zosimos (1938/1954)
Paracelsus as a Spiritual Phenomenon (1942)
The Spirit Mercurius (1943/1948)
The Philosophical Tree (1945/1954)

14. MYSTERIUM CONIUNCTIONIS ([1955–56] 1963; 2d ed., 1970)

AN INQUIRY INTO THE SEPARATION AND SYNTHESIS OF PSYCHIC OPPOSITES IN ALCHEMY

The Components of the Coniunctio
The Paradoxa
The Personification of the Opposites
Rex and Regina
Adam and Eve
The Conjunction

15. THE SPIRIT IN MAN, ART, AND LITERATURE (1966)

Paracelsus (1929)
Paracelsus the Physician (1941)
Sigmund Freud in His Historical Setting (1932)
In Memory of Sigmund Freud (1939)
Richard Wilhelm: In Memoriam (1930)
On the Relation of Analytical Psychology to Poetry (1922)
Psychology and Literature (1930/1950)
"Ulysses": A Monologue (1932) Picasso (1932)

16. THE PRACTICE OF PSYCHOTHERAPY (1954; 2d ed., 1966)

GENERAL PROBLEMS OF PSYCHOTHERAPY

Principles of Practical Psychotherapy (1935)
What is Psychotherapy? (1935)
Some Aspects of Modern Psychotherapy (1930)
The Aims of Psychotherapy (1931)
Problems of Modern Psychotherapy (1929)
Psychotherapy and a Philosophy of Life (1943)
Medicine and Psychotherapy (1945)
Psychotherapy Today (1945)
Fundamental Questions of Psychotherapy (1951)

SPECIFIC PROBLEMS OF PSYCHOTHERAPY
The Therapeutic Value of Abreaction (1921/1928)
The Practical Use of Dream-Analysis (1934)
The Psychology of the Transference (1946)
Appendix: The Realities of Practical Psychotherapy ([1937] added 1966)

17. THE DEVELOPMENT OF PERSONALITY (1954)
Psychic Conflicts in a Child (1910/1946)
Introduction to Wickes's "Analyses der Kinderseele" (1927/1931)
Child Development and Education (1928)
Analytical Psychology and Education: Three Lectures (1926/1946)
The Gifted Child (1943)
The Significance of the Unconscious in Individual Education (1928)
The Development of Personality (1934)
Marriage as a Psychological Relationship (1925)

18. THE SYMBOLIC LIFE (1954)
Translated by R.F.C. Hull and others
Miscellaneous Writings

19. COMPLETE BIBLIOGRAPHY OF C. G. JUNG'S WRITINGS (1976; 2d ed., 1992)

20. GENERAL INDEX OF THE COLLECTED WORKS (1979)

THE ZOFINGIA LECTURES (1983)
Supplementary Volume A to the Collected Works.
Edited by William McGuire, translated by
Jan van Heurck, introduction by
Marie-Louise von Franz

PSYCHOLOGY OF THE UNCONSCIOUS ([1912] 1992)
A STUDY OF THE TRANSFORMATIONS AND SYMBOLISMS OF THE LIBIDO.
A CONTRIBUTION TO THE HISTORY OF THE EVOLUTION OF THOUGHT
Supplementary Volume B to the Collected Works.
Translated by Beatrice M. Hinkle,
introduction by William McGuire

Notes to C. G. Jung's Seminars

DREAM ANALYSIS ([1928–30] 1984)
Edited by William McGuire

NIETZSCHE'S *ZARATHUSTRA* ([1934–39] 1988)
Edited by James L. Jarrett (2 vols.)

ANALYTICAL PSYCHOLOGY ([1925] 1989)
Edited by William McGuire

THE PSYCHOLOGY OF KUNDALINI YOGA ([1932] 1996)
Edited by Sonu Shamdasani

INTERPRETATION OF VISIONS ([1930–34] 1997)
Edited by Claire Douglas

Philemon Series of the Philemon Foundation

Genaral editor, Sonu Shamdasani

Children's Dreams. Edited by Lorenz Jung and Maria Meyer-Grass. Translated by Ernst Falzeder with the collaboration of Tony Woolfson

Introduction to Jungian Psychology: Notes of the Seminar on Analytical Psychology Given in 1925. Edited by William McGuire. Translated by R.F.C. Hull. With a new introduction and updates by Sonu Shamdasani

Jung contra Freud: The 1912 New York Lectures on the Theory of Psychoanalysis. With a new introduction by Sonu Shamdasani. Translated by R.F.C. Hull

The Question of Psychological Types: The Correspondence of C. G. Jung and Hans Schmid-Guisan, 1915–1916. Edited by John Beebe and Ernst Falzeder. Translated by Ernst Falzeder with the collaboration of Tony Woolfson

Dream Interpretation Ancient and Modern: Notes from the Seminar Given in 1936–1941. C. G. Jung. Edited by John Peck, Lorenz Jung, and Maria Meyer-Grass. Translated by Ernst Falzeder with the collaboration of Tony Woolfson

Analytical Psychology in Exile: The Correspondence of C. G. Jung and Erich Neumann. Edited and introduced by Martin Liebscher. Translated by Heather McCartney

On Psychological and Visionary Art: Notes from C. G. Jung's Lecture on Gérard de Nerval's "Aurélia." Edited by Craig E. Stephenson. Translated by R.F.C. Hull, Gottwalt Pankow, and Richard Sieburth